AGRICULTURAL ISSUES AND POLICIES

SUGAR: BACKGROUND, POLICIES AND ISSUES

AGRICULTURAL ISSUES AND POLICIES

Additional books in this series can be found on Nova's website under the Series tab.

Additional E-books in this series can be found on Nova's website under the E-books tab.

AGRICULTURAL ISSUES AND POLICIES

SUGAR: BACKGROUND, POLICIES AND ISSUES

CECILIA M JORDAN
EDITOR

Nova Science Publishers, Inc.
New York

For permission to use material from this book please contact us:
Telephone 631-231-7269; Fax 631-231-8175
Web Site: http://www.novapublishers.com

LIBRARY OF CONGRESS CATALOGING-IN-PUBLICATION DATA

Sugar : background, policies and issues / editor: Cecilia M. Jordan.
p. cm.
Includes index.
ISBN 978-1-61761-116-2 (hardcover)
1. Sugar trade--Government policy. 2. Agricultural price supports. 3. Sugar trade--Statistics. 4. Sweetener industry--Statistics. I. Jordan, Cecilia M.
HD9100.6.S84 2010
338.4'76641--dc22
2010027153

Published by Nova Science Publishers, Inc. ✢ New York

CONTENTS

PREFACE

Governments of every sugar-producing nation intervene to protect their domestic industries from fluctuating world market prices. Such intervention is necessary, it is argued, because both sugar cane and sugar beets must be processed soon after harvest using costly processing machinery. When farmers significantly reduce production because of low prices, a cane or beet processing plant typically shuts down, usually never to reopen. This close link between production and capital-intensive processing makes price stability important to industry survival. This book examines the background, policies and issues of the sugar industry across the globe.

Chapter 1- The sugar program, as authorized by the 2008 farm bill (P.L. 110-246), is designed to guarantee that growers and processors of sugar beets and sugarcane receive a minimum price. To do this, the U.S. Department of Agriculture (USDA), in projecting domestic use of sugar for food, limits the amount of domestically produced sugar that processors can sell under "marketing allotments" and restricts imports. These decisions to control supply are aimed at keeping market prices above support levels, so that USDA can operate the sugar program at "no cost" to the U.S. Treasury. Separately, trade law authorizes the Secretary of Agriculture to allow additional imports if the domestic sugar supply is not adequate to meet domestic demand at reasonable prices.

The United States is the fifth-largest consumer of sugar in the world. Consumption in food has increased in recent years, reflecting population growth and a shift back from the use of high- fructose corn syrup, an alternative and cheaper sweetener. Adverse weather significantly reduced beet sugar output in FY2009 and contributed to tight sugar supplies as the current marketing year began. USDA projects an even tighter picture by the end of FY20 10, with ending stocks relative to demand expected to reach a 36-year low. As a result, the raw cane sugar futures price and the refined beet sugar spot price have increased by 64% and 36%, respectively, from July 2009 to mid-January 2010. These prices are considerably above their support levels and 10-year averages.

Against this backdrop, industrial sugar users (food and beverage manufacturers) have called on USDA to allow for additional sugar imports to head off sugar "shortages" and to restore supplies to more normal levels. Sugar crop growers and processors maintain that the domestic market is adequately supplied, with the harvesting and processing of the 2009 crops well underway. Though these contrasting views spotlight the matter of sugar availability, the underlying issue has more to do with the future level of the price of sugar. Sugar processors seek to maintain their advantage in negotiating higher sales prices with users. Food and beverage firms want refined sugar prices that are lower and closer to the historical average.

Chapter 2- Congress reauthorized the sugar price support program with some changes in the Food, Conservation, and Energy Act of 2008 (P.L. 110-246, the enacted 2008 farm bill). The sugar program is designed to guarantee the price received by sugar crop growers and processors and is intended to operate at "no cost" to the U.S. Treasury. To accomplish this, the U.S. Department of Agriculture (USDA) controls supply by limiting the amount of sugar that processors can sell domestically under "marketing allotments" and restricts imports. At the same time, USDA seeks to ensure that supplies of sugar are adequate to meet domestic demand. "No cost" is achieved if USDA applies these tools in a way that maintains market prices above minimum price support levels.

Since January 1, 2008, sugar imports from Mexico no longer face quotas or duties under the North American Free Trade Agreement. Other imports are allowed entry under quotas found in other free trade agreements (FTAs). To address the potential for a U.S. sugar surplus caused by additional imports under these trade agreements, the enacted farm bill mandates a sugar-forethanol program. USDA is now required to purchase as much U.S.-produced sugar as necessary to maintain market prices above support levels, to be sold to bioenergy producers for processing into ethanol. Funding is open-ended for this program. Other provisions increase the minimum guaranteed prices for raw sugar and refined beet sugar by 4% to 5%, mandate an 85% market share for the U.S. sugar production sector, and remove certain discretionary authority that USDA exercises to administer import quotas.

The enacted sugar provisions reflect the proposal presented to the House and Senate Agriculture Committees by producers of sugar beets and sugarcane and the processors of these crops. They favored continuing the structure of the current sugar price support program but sought changes to enhance their position in the U.S. marketplace. Their sugar-for-ethanol provisions ensure that the prospect of imports adding to U.S. sugar supplies under any future trade agreements will not undermine the program's price guarantee and the sugar industry's market share. Food and beverage manufacturers that use sugar opposed the proposed program's provisions, arguing that costs to consumers will increase and that new requirements will restrict the flow of sugar for food use in the domestic market. The Bush Administration opposed these provisions, with the President identifying them as one reason why he vetoed the farm bill.

Chapter 3- A cyclical decline in sugar production is shifting India, the world's second largest producer, from net exporter to net importer during 2009/10 (October/September) and contributing to the current runup in global sugar prices. The downturn in Indian production is primarily due to a policy-induced cycle that has become increasingly pronounced. Large domestic surpluses during 2006/07 and 2007/08 led to lower domestic sugar prices, weakened incen-tives for sugarcane growers, and sharply reduced plantings for the 2008/09 and 2009/10 crops. Indian domestic sugar prices have now increased sharply since early 2009 because of reduced supplies—and despite large government releases of sugar and liquidation of government-owned buffer stocks. India is forecast to shift from net sugar exports of 5.8 million tons in 2007/08 to net imports of 2.8 million tons in 2008/09 and a record 6.0 million tons in 2009/10.

Indian sugarcane and sugar production are poised to rebound in 2010/11. A combination of higher government price supports and open-market prices will likely stimulate plantings and also improve incentives to deliver sugarcane to sugar mills rather than to alternate uses.

In the longer term, India has the capacity to boost both area and yields of sugarcane, and the government and the sugar industry are engaged in an ongoing dialogue on policy measures to moderate the increasingly sharp cycles in sugar production and trade.

Chapter 4- The European Union (EU) is one of the leading sugar producers and traders in the world. This position was built over time through the application of protectionist policies that regulated production, prices, exports, and imports. Since its creation in 1968, the existing EU sugar policy—commonly referred to as the Common Market Organization (CMO) for sugar—had changed only marginally. In 2006, however, a new sugar regime took effect, largely influenced by three factors: tariff- and quota-free access to least developed countries (LDCs) beginning in 2009, which would likely result in increased sugar imports; enforcement of World Trade Organization (WTO) commitments; and the accession of 10 new member states to the EU in 2004, which exacerbated preexisting sugar supply-and-demand imbalances.

Chapter 5- U.S. sugar supply for fiscal year (FY) 2011 is projected to be 3.3 percent less than in FY 2010. Although production in FY 2011 is expected to be higher than in FY 2010 by 318,000 short tons, raw value (STRV), lower beginning stocks (269,000 STRV less) and lower imports (435,000 STRV less) are more than offsetting. Higher beet sugar production reflects a return to trend yields, while cane sugar production is increased for Florida and Texas. Imports under the tariff-rate quota (TRQ) are projected at the minimum of U.S. commitments to import raw and refined sugar, including an expected shortfall. The Secretary of Agriculture will establish the actual level of the TRQ at a later date. Imports from Mexico are nearly unchanged. Total use (combined deliveries, exports, and refining losses) is unchanged.

Mexico's FY 2011 sugar supply is up 1.7 percent, with higher stocks and production more than offsetting lower imports. Production is projected to increase, as yields rebound to trend levels. Imports reflect mainly U.S. exports for Mexico's IMMEX program. Domestic sugar consumption is down slightly, due to higher expected use of corn-based sweeteners, and exports are up slightly. Ending stocks are moderately above a year earlier.

For FY 2010 U.S. sugar, increased supplies nearly offset increased use, compared with a month earlier. Imports are increased under the TRQ, in line with USDA's April 23, 2010, announcement and expectations for increased imports at the high-tier tariff. Cane sugar production is reduced. Sugar deliveries are increased to reflect the recent strong pace to date and improved data availability that made possible better analysis of recent trends.

Chapter 6- In the February 2010 *World Agriculture Supply and Demand Estimates* (WASDE), projected fiscal year (FY) 2010 production for Mexico is reduced 200,000 metric tons, raw value (MTRV) from last month based on weather-reduced sugar yields to date. Exports are reduced by the same amount. Projected FY 2010 U.S. sugar supply is decreased by 85,000 short tons, raw value (STRV) from last month due to lower imports from Mexico, more than offsetting higher sugar production. Imports from Mexico are reduced by 220,000 STRV. Beet sugar production and Louisiana cane sugar production are increased a combined 150,000 STRV due to anticipated output in September 2010 from next season's sugar crops. Florida cane sugar production is reduced 15,000 STRV based on a lower forecast of sugarcane for this year by the National Agricultural Statistics Service (NASS). Total use is unchanged. Ending-year stocks are projected at 1.055 million STRV, implying an ending stocks-to-use ratio of 10.0 percent.

Chapter 7- The National Agricultural Statistics Service (NASS) forecasts sugar beet acreage intentions for crop year 2009/10 at 1.152 million acres, about 5.7 percent higher than area planted in crop year 2008/09. Assuming normal sucrose levels, normal area harvested-to-planted ratios, and continued improvement in productivity, the U.S. Department of Agriculture (USDA) projects national beet sugar production in crop year 2009/10 at 4.610 million short tons, raw value (STRV). Because 60,000 STRV of this production is expected to occur in September (the last month of fiscal year (FY) 2009), FY 2010 beet sugar production is projected at 4.550 million STRV, according to the *World Agricultural Supply and Demand Estimates* (WASDE) report. This amount is about 6.8 percent higher than the estimated 4.260 million STRV in FY 2009.

The USDA projects FY 2010 cane sugar production at 3.525 million STRV, an increase of 179,000 STRV over that of FY 2009. Because area harvested is not forecast by NASS until the end of June, current cane sugar projections assume the same area harvested for sugar as in the previous year. Florida cane sugar production for FY 2010 is forecast at 1.750 million STRV, assuming the continued dry conditions that have reduced output in the last three growing seasons. The USDA projects FY 2009 Louisiana cane sugar production at 1.400 million STRV. However, hurricane damage in fall of 2008 caused delays in sugarcane planting that could affect the size of the crop. The USDA projects Texas production at 165,000 STRV and Hawaiian production at 210,000 STRV.

Although the raw and refined sugar tariff-rate quotas (TRQ) for FY 2010 have not yet been announced, the USDA projects them at minimum levels implied by existing international commitments to the World Trade Organization (WTO) and at the allocated levels of existing Free Trade Agreements (FTAs). The projection in the May WASDE is, therefore, at 1.232 million STRV. Projected shortfall at 150,000 STRV is at about the same higher-than-normal level estimated in FY 2009 (155,000 STRV) because several quota-holding countries are diverting sugar to the European Union (EU) before preferential prices available to them are reduced according to the EU's sugar reform schedule.

In: Sugar: Background, Policies and Issues
Editor: Cecilia M. Jordan
ISBN: 978-1-61761-116-2

Chapter 1

SUGAR POLICY ISSUES*

Remy Jurenas

SUMMARY

The sugar program, as authorized by the 2008 farm bill (P.L. 110-246), is designed to guarantee that growers and processors of sugar beets and sugarcane receive a minimum price. To do this, the U.S. Department of Agriculture (USDA), in projecting domestic use of sugar for food, limits the amount of domestically produced sugar that processors can sell under "marketing allotments" and restricts imports. These decisions to control supply are aimed at keeping market prices above support levels, so that USDA can operate the sugar program at "no cost" to the U.S. Treasury. Separately, trade law authorizes the Secretary of Agriculture to allow additional imports if the domestic sugar supply is not adequate to meet domestic demand at reasonable prices.

The United States is the fifth-largest consumer of sugar in the world. Consumption in food has increased in recent years, reflecting population growth and a shift back from the use of high- fructose corn syrup, an alternative and cheaper sweetener. Adverse weather significantly reduced beet sugar output in FY2009 and contributed to tight sugar supplies as the current marketing year began. USDA projects an even tighter picture by the end of FY20 10, with ending stocks relative to demand expected to reach a 36-year low. As a result, the raw cane sugar futures price and the refined beet sugar spot price have increased by 64% and 36%, respectively, from July 2009 to mid-January 2010. These prices are considerably above their support levels and 10-year averages.

Against this backdrop, industrial sugar users (food and beverage manufacturers) have called on USDA to allow for additional sugar imports to head off sugar "shortages" and to restore supplies to more normal levels. Sugar crop growers and processors maintain that the domestic market is adequately supplied, with the harvesting and processing of the 2009 crops well underway. Though these contrasting views spotlight the matter of sugar availability, the

* This is an edited, reformatted and augmented edition of a United States Congressional Research Service publication, Report R40995, dated January 19, 2010,

underlying issue has more to do with the future level of the price of sugar. Sugar processors seek to maintain their advantage in negotiating higher sales prices with users. Food and beverage firms want refined sugar prices that are lower and closer to the historical average.

The debate over additional sugar imports centers around a provision in the 2008 farm bill that prescribes USDA's authority in administering the import quotas established to meet U.S. commitments under the World Trade Organization. It requires USDA at the beginning of each fiscal year to set the quotas for raw sugar and refined sugar at the minimum levels laid out in this obligation. Imports can be increased before April 1 only if there is an "emergency shortage of sugar" due to war or a natural disaster. USDA acknowledges that additional sugar supplies will be needed, and plans to monitor market developments. Sugar users continue to call for a USDA decision before April 2010. Processors want USDA to wait. Members of Congress have weighed in with letters to USDA in support of each side's position. The timing of a decision, if one occurs, will continue to affect near-term refined sugar prices and influence negotiations between processors and users on the price of sugar to be delivered in the future.

Free trade in sweeteners with Mexico since 2008 has introduced considerable uncertainty as to how much sugar Mexico might export in any year to the U.S. market. Actual sales have varied considerably, influenced largely by whether U.S. prices are higher than those in Mexico. This, in turn, complicates USDA's effort to administer the sugar program, particularly to ensure that market prices remain above support levels and that the statutory no-cost objective is met. To address this uncertainty, sugar processors advocate increased coordination between the U.S. and Mexican governments on sugar policies. Sugar users oppose this "managed-trade proposal," arguing that it would result in inadequate domestic supplies and hurt U.S. jobs.

Background on the Sugar Program

Governments of every sugar-producing nation intervene to protect their domestic industries from fluctuating world market prices. Such intervention is necessary, it is argued, because both sugar cane and sugar beets must be processed soon after harvest using costly processing machinery. When farmers significantly reduce production because of low prices, a cane or beet processing plant typically shuts down, usually never to reopen. This close link between production and capital-intensive processing makes price stability important to industry survival.

The United States has a long history of protection and support for its sugar industry. The Sugar Acts of 1934, 1937, and 1948 required the U.S. Department of Agriculture (USDA) to estimate domestic sugar consumption and divide this market by assigning quotas to U.S. growers and foreign countries. These acts also authorized payments to growers when needed as an incentive to limit production and levied excise taxes on sugar processed and refined in the United States. This type of sugar program expired in 1974. For the next seven years, the U.S. market was relatively open to foreign sugar imports, with mandatory price support provided only in 1977 and 1978, and discretionary support in 1979. Congress reinstated mandatory price support for sugar in the Agriculture and Food Act of 1981 and the Food Security Act of 1985. Subsequently, the 1990 farm bill, the 1993 budget reconciliation bill, and the 1996, 2002, and 2008 farm bills extended sugar program authority with some changes. Current law authorizes the sugar program through the 2012 crops for sugar beets and sugarcane.

Even with price protection available to producers, the United States historically has not produced enough sugar to satisfy domestic demand and thus continues to be a net sugar importer. Prior to the early 1980s, domestic sugar growers supplied roughly 55% of the U.S. sugar market. This share has grown over the last 25 years, reflecting the price protection provided by the sugar program. As high-fructose corn syrup (HFCS) displaced sugar in the United States during the early 1980s, and domestic sugar production increased later in that decade, foreign suppliers absorbed the entire adjustment and saw their share of the U.S. market decline significantly.

Current U.S. sugar policy maintains domestic sugar prices above the world market price, and is structured to protect the domestic sugar-producing sector (sugar beet and sugarcane producers, and the processors of their crops) and to ensure a sufficient supply. During the last decade (FY2000-FY2009), the U.S. raw sugar domestic futures price averaged 21.0 cents per pound (¢/lb.), compared to the world raw sugar futures price of 10.8¢/lb. Because of the price differential, U.S. consumers and manufacturers of foods and beverages pay more for sugar than they would if imports were allowed to enter without any restriction. Various studies show that over the last 15 years, U.S. sugar users paid between $400 million and $1.9 billion more for sugar annually. These "cost to user" estimates vary widely, and largely reflect the extent of the difference between the higher U.S. price and the lower world price for sugar in the time period examined, and the differing assumptions and methodologies that analysts use to develop such estimates.

The sugar program differs from the grains, rice, peanut, and cotton programs in that USDA makes no direct payments to beet and cane growers and processors. With this structure, taxpayers do not directly support the program through federal government outlays. This fact is highlighted as a positive feature by the sugar production sector and program supporters. The program's support level and import protection, however, keep the U.S. sugar price above the price of sugar traded internationally, and constitute an indirect subsidy to the production sector by way of higher prices paid by U.S. sugar users (food and beverage manufacturers) and household consumers. Program opponents frequently refer to this subsidy component to argue for changes to U.S. sugar policy.

MAIN FEATURES OF U.S. SUGAR POLICY

U.S. sugar policy uses three tools to ensure that domestic growers and sugar processors receive a minimum price for their sugar. First, statutorily set loan rates determine the support level that processors, and in turn producers, are guaranteed to receive for raw cane and refined beet sugar. Second, "marketing allotments" limit the amount of domestically produced sugar that processors can sell, and now ensure that the domestic sugar production sector supplies at least 85% of U.S. food and beverage demand for sugar. Third, import quotas restrict the amount of foreign sugar allowed to enter the U.S. market. USDA decisions in administering allotments and quotas are intended to balance available sugar supply (i.e., domestic output plus imports) with U.S. food demand for sugar so that market prices do not fall below effective support levels. The 2008 farm bill (P.L. 110-246, Subtitle D of Title I) further requires USDA to operate the sugar program on a "no-cost" basis (i.e., to result in no federal government outlays). It is administered by the Farm Service Agency. Each tool is described below, with a discussion of what the 2008 farm bill requires and what USDA has announced in exercising its authorities for the FY2009 and FY2010 sugar programs.[1]

Price Support Loans

USDA extends "non-recourse" price support loans to processors of sugarcane and sugar beets rather than directly to the farmers who harvest these crops.[2] Loans are available only to processors who agree to pay growers for deliveries of sugar beets and sugarcane at USDA-set minimum payment levels. Their "non-recourse" feature means that a processor can exercise the legal right to hand over sugar it initially offered USDA as collateral for the loan to meet its repayment obligation, if the market price is below the effective support level when the loan comes due. These loans at times can be attractive to sugar processors as a source of short-term credit at below-prime interest rates.

Loan Rates

The 2008 farm bill increases sugar loan rates by 4% to 5% in stages through FY2012. For the 2009 sugarcane and sugar beet crops that are harvested and processed during FY2010, the national average loan rate is 18.25 ¢/lb. for raw cane sugar, and 23.45 ¢/lb. for refined beet sugar. The loan rate increase is the first since 1985 for raw cane sugar and 1992 for refined beet sugar. The rate for beet sugar is statutorily set higher than for raw sugar because the former is available immediately after processing in white refined form, ready for industrial food and beverage use and for human consumption. By contrast, raw cane sugar must go through a second stage of processing at a cane refinery to be converted into white refined sugar. To reflect differences in marketing costs among production areas, USDA adjusts these national loan rates on a regional basis.[3]

Effective Support Levels

Sugar loan rates alone do not serve as the guaranteed price floor for domestically produced sugar. In practice, USDA's aim is to support raw cane and refined beet sugar at higher price levels that take into account those costs that a processor incurs in taking out a price support loan and in marketing sugar to a first purchaser. During FY2009, the effective support level for raw cane sugar was 19.81 ¢/lb. in Louisiana and Texas, and about 20.6 ¢/lb. in Florida and Hawaii. These were based on each state's price support level, *plus* an amount to cover a processor's cost to transport raw cane sugar to a cane refinery, *plus* the interest paid on any price support loan taken out, *plus* location discounts. For refined beet sugar, effective support across six beet-producing regions ranged from about 23 ¢/lb. to about 26 ¢/lb. The support level was based upon the regional loan rate, *plus* specified marketing costs, *plus* interest paid on a price support loan, *plus* a cash discount. For FY20 10, effective support levels are expected to stay about the same.

To ensure that market prices do not fall below these "effective" price support levels, USDA limits the amount of sugar that can be marketed by administering marketing allotments and import quotas (i.e., limiting supply). A loan forfeiture (turning over sugar pledged as loan collateral to USDA) occurs if a processor concludes that the domestic market price when the loan comes due is below the "effective" sugar support level for his state or region. In other words, this is the price that would discourage a processor from "forfeiting" sugar placed under loan to USDA. It is this level—not the loan rate—that represents the minimum level of program benefits that sugar crop growers and processors expect to receive.

Market prices ended FY2009 (the first year that the 2008 farm bill-authorized sugar program was in effect) significantly above loan forfeiture levels. Though the raw cane sugar futures price for the July-September 2009 period, when price support loans normally come due, skirted around the loan forfeiture range for much of the first half of the marketing year (Figure 1), prices began to rise in April 2009 once it became clear that USDA would not increase raw cane sugar imports. By contrast, refined beet sugar prices were considerably above loan forfeiture levels during all of FY2009. Prices began to rise in August 2009 once USDA signaled that there would be no increase in the import quota for refined sugar before September 30 (Figure 2). Various explanations for low raw sugar and high refined beet prices during FY2009 were offered. Some pointed out that continued imports of raw-sugar-equivalent product from Mexico that needed additional refining kept downward pressure on the U.S. raw cane sugar futures price. Others noted that the loss of the production capacity associated with a major cane sugar refinery, caused by an explosion in February 2008, reduced U.S. demand for raw cane sugar—its primary feedstock—and kept the raw cane sugar price lower than might be the case otherwise. Higherthan-average refined sugar prices are also attributed to the loss of refined product that this refinery would normally have supplied.[4]

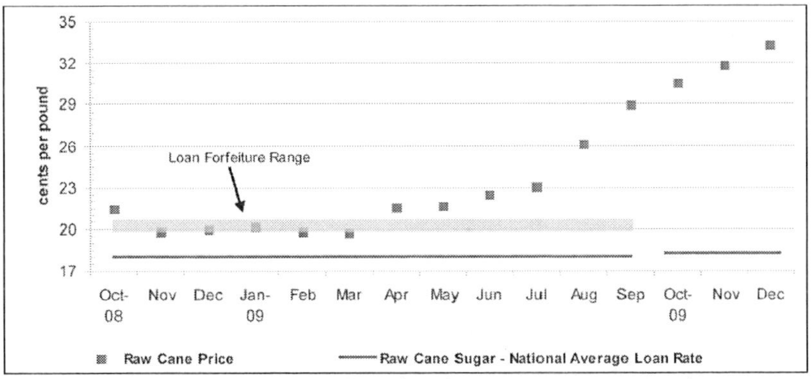

Source: USDA, Economic Research Service, for price data; USDA, Farm Service Agency, for FY2009 loan forfeiture range.

Note: Raw cane market price is the average futures price for the nearby month contract for #16, traded in New York City on the IntercontinentalExchange (ICE).

Figure 1. Raw Cane Sugar Prices Fell Within and Below Loan Forfeiture Range in First Half of 2008/2009 Marketing Year.

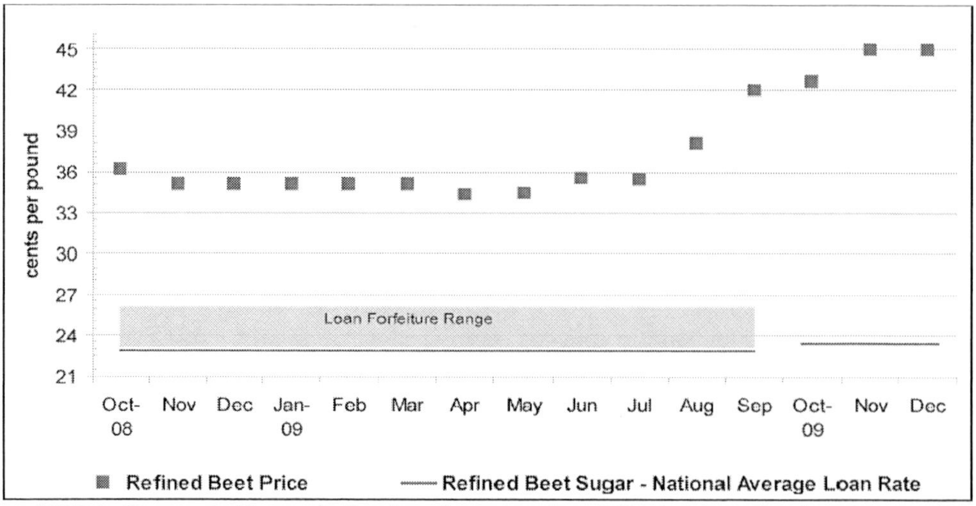

Source: USDA, Economic Research Service, for price data; USDA, Farm Service Agency for loan forfeiture range (using FY2008 levels as proxy for FY2009).

Notes: Market price for refined beet sugar are the quotations for wholesale refined beet sugar, Midwest markets, as published weekly by *Milling and Baking News*.

Figure 2. Refined Beet Sugar Prices Higher than Loan Forfeiture Range During Entire 2008/2009 Marketing Year.

Marketing Allotments

In the 2002 farm bill, the domestic production sector accepted mandatory limits on the amount of sugar that processors can sell—known as marketing allotments—in return for the assurance of price protection. The sector viewed allotments as a way to try to capture any growth in U.S. sugar demand, and assumed that the then-U.S. sugar import quota commitments would continue without change through the end of the farm bill period (September 30, 2008).[5] However, concerned that their market share would decline as sugar imports increase under various free trade agreements (in particular from Mexico beginning in 2008), sugar producers and processors secured a significant change in the 2008 farm bill on how the national marketing allotment is to be determined. USDA is now required each year to set an "overall allotment quantity" (OAQ)— the amount of sugar that all processors combined can sell—that is not less than 85% of estimated U.S. sugar consumption for food. This provision guarantees a minimum 85% market share for the domestic sector, and is intended to ensure that imports do not displace the ability of U.S. sugar processors to sell more of their output in each successive year, to the extent that there is growth in U.S. demand for sugar.

This change removed discretionary authority that USDA previously exercised, when it weighed four specified factors in determining the OAQ, or the amount of domestic sugar allowed to be marketed. Then, sugar processors and food manufacturers (i.e., users of sugar) weighed in to influence USDA's decision-making process on this issue. Each group differed in how USDA should define "reasonable" ending stocks—a key determinant of the level of domestic sugar prices in the last quarter of the marketing year (July to September) when price support loans come due. Sugar processors favored a smaller OAQ, seeking to benefit from sugar prices well above effective support levels. Food manufacturers advocated a larger OAQ, hoping that year-end prices would end up lower, and close to loan forfeiture levels. With USDA now only required to set the OAQ at not less than 85% of projected sugar food use, interest group focus has turned toward influencing USDA decisions on import quotas, where its discretionary authority is now considerably prescribed (see "2008 Farm Bill Sugar TRQ Provision," below).

The 2008 farm bill continues to split the OAQ between the beet and cane sectors using 54.35% and 45.65% shares, respectively. Additional rules specify how each sector's allotment is to be allocated (i.e., distributed) to each sugar crop processing firm. Once USDA makes detailed calculations using specified criteria, each firm can sell only as much sugar as stated in its allocation notification received from USDA. Sugar produced and held in inventory in excess of a firm's allocation must be held off the market (referred to as "blocked stocks"). However, if a beet or cane processing firm is not able to fill its sugar allocation, USDA can reassign the sum of all such "shortfalls" to raw cane sugar imports.

For FY2009, USDA set the final OAQ at 9.235 million short tons (ST), raw value[6]—equal to 85% of U.S. food consumption of sugar as estimated in August 2009. Because this quantity was 11% higher than what all firms reported they had available to market (production from processing the 2008 crops, plus stocks held in inventory), USDA assigned a shortfall of 1.0 million ST to imports of raw cane sugar.[7] A large portion of the sugar imported from Mexico can be viewed as having covered this shortfall. For FY2010, USDA set the initial OAQ at 9.235 million ST,[8] or 91% of its September 2009 estimate of food sugar consumption for the year. Because raw cane and beet sugar output then was projected to be 13% below the announced OAQ, USDA expects domestic processors again to sell all of their production. It also noted that since raw cane sugar output "is expected to fall significantly short of its allotment," this shortfall will be reassigned to imports later in FY2010.[9] Though the domestic supply outlook can change during a marketing year, an allotments shortfall of about 1.0 million ST, reassigned to imports, is possible again this year.

Import Quotas

The United States imports sugar to cover the balance of the demand that the U.S. production sector cannot supply. In FY2009, imports accounted for 26% of the sugar consumed domestically for food. USDA, however, restricts the quantity of foreign sugar allowed to enter for refining and/or direct sale to manufacturers for domestic food and beverage use. Tariff-rate quotas (TRQs) are used to restrict sugar imports to the extent needed to meet U.S. sugar program objectives.[10] In other words, they serve to ensure that the amounts allowed to enter do not depress the domestic market price to below effective support levels, and that USDA as a result does not acquire sugar due to processors' loan forfeitures (i.e., that USDA is able to meet the "no-cost" requirement). Quota amounts are laid out in U.S. market access commitments made under World Trade Organization (WTO) rules and under bilateral free trade agreements (FTAs).

The U.S. market access commitment made under WTO rules means that a minimum of 1.256 million ST of foreign sugar must be allowed to enter the U.S. market each year.

Although the WTO commitment sets this minimum level, the Secretary of Agriculture has authority to allow additional amounts of sugar to enter if "domestic supplies of sugars may be inadequate to meet domestic demand at reasonable prices."[11] The U.S. Trade Representative (USTR) allocates the 1.231 million ST raw cane sugar TRQ among 40 eligible countries. The much smaller refined sugar TRQ is allocated first to Canada and Mexico, with the balance reserved for the rest of the world on a first-come, first-served basis (Table 1).[12]

Table 1. U.S. Market Access Quota Commitments for Sugar Imports (short tons, raw value).

Trade Agreement & Quota Subcategory	2008[a]	2009[a]	2010[a]
WTO TRQ			
Raw Cane Sugar[b]	1,231,497	1,231,497	1,231,497
Refined Sugar[c]	394,250[d]	104,251	99,251
Subtotal	1,625,747	1,335,748	1,330,748
Mexico/NAFTA			
Sugar[e]	694,000	1,402,000	760,000
DR-CAFTA TRQ			
Sugar & Sugar Goods	98,591	119,557	119,163
Specialty Sugar[f]	—	2,205	2,205
Subtotal	98,591	121,761	121,368
Peru FTA TRQ			
Sugar & Sugar Goods[g]	—	0	0
Specialty Sugar	—	2,205	2,205
TOTAL	2,418,338	2,861,714	2,214,339

Source: Office of the U.S. Trade Representative, sugar TRQ press release, September 26, 2009, notice published in *Federal Register*, December 16, 2009, p. 66718; USDA, World Agricultural Outlook Board, *World Agricultural Supply and Demand Estimates*, January 12, 2010; U.S. International Trade Commission, *Harmonized Tariff Schedule of the United States*, Chapter 17 (WTO), Chapter 98 (DR-CAFTA and Peru FTA).

Notes: Quantities specified in trade agreements here are converted into English units (i.e., metric tons to short tons). One metric ton equals 1.10231125 short tons (2,000 pounds). All TRQs above reflect duty-free access for the quantities shown, except for three countries under the WTO raw cane sugar TRQ and several countries under the WTO refined sugar TRQ. See Figure 10 for a graphic overview of actual sugar imports under various trade agreements.

a. Fiscal year for WTO and Mexico/NAFTA; calendar year for DR-CAFTA and Peru FTA.

b. Announced at minimum commitment level (equal to 1,117,195 metric tons (MT)) for each year shown.

c. Minimum commitment is 22,000 MT (or 24,251 ST), but USDA each year has allowed additional entry for specialty sugars to meet demand by the rapidly expanding organic food sector (on average about 75,000 ST annually).

d. USDA increased the refined sugar TRQ late in FY2008 by 272,155 MT (300,000 ST) in response to a tight market resulting from the loss of refining capacity at one cane sugar refinery and a reduction in expected sugar beet output from the 2008 crop.

e. Under NAFTA, Mexico received access to export unlimited amounts of sugar duty-free to the U.S. market, beginning on January 1, 2008. Quantities shown are recorded U.S. imports for FY2008 and FY2009, and USDA-projected imports for FY20 10. Current year (FY20 10) imports from Mexico likely will change to reflect evolving market conditions in both the U.S. and Mexican sweetener markets (see "Free Trade under NAFTA").

f. Available only to Costa Rica, beginning January 1, 2009.

g. Under the agreement's terms (which took effect on February 1, 2009), Peru can use its preferential sugar TRQ only if it shows a sugar trade surplus. Since USTR calculated that this was not the case in 2007 and 2008, Peru did not have additional access under this FTA in 2009, nor will it in 2010.

In a significant policy change, the 2008 farm bill requires USDA to set the WTO sugar TRQ commitments at the minimum levels at the beginning of the quota (i.e., marketing) year, lays out the factors to be considered to increase quota imports if there is an emergency sugar shortage before April 1 of each year, details what import actions USDA can take before and after April 1, and prescribes the mix of raw versus refined sugar allowed to enter, depending upon which half of the year such entries occur (see "2008 Farm Bill Sugar TRQ Provision").

Effective January 1, 2008, the end of the North American Free Trade Agreement's (NAFTA) transition period to free trade, Mexico no longer faces any tariff or quantitative limit on the amount of sugar that can be exported to the U.S. market. With this opening, imports have fluctuated considerably from year to year (almost 700,000 ST in FY2008, 1.4 million ST in FY2009, and 760,000 ST projected in FY20 10; see Table 1). The doubling of Mexican exports in FY2009 over the previous year appears to be largely in response to higher prices for refined sugar in the U.S. market than mills could receive in Mexico. See "Free Trade under NAFTA" for discussion of implications of this year's Mexican exports on FY20 10 U. S. supply and prices.

Under other bilateral FTAs, the United States has committed to allow entry to additional sugar imports from partners that already receive allocations under the U.S. WTO raw cane sugar TRQ. The most significant is the preferential access granted under the Dominican Republic-Central American FTA (DR-CAFTA) to Costa Rica, Dominican Republic, El Salvador, Guatemala, Honduras, and Nicaragua. In 2010, five of these countries are eligible to ship up to an additional 124,000 ST of sugar to the U.S. market (Table 1). Their preferential quotas combined will increase on average almost 3% per year, and total almost 169,000 ST in 2020 (year 15 of the agreement). Thereafter, quotas in the aggregate will increase by 1.7% (2,910 ST) annually in perpetuity. Though other FTAs grant other country partners formal access for their sugar to the U.S. market, actual exports will be minimal because these partners are not expected to show that they have a surplus in their sugar trade with the rest of the world.[13]

Under all trade agreements, the amount of sugar that enters under the "in-quota" portion of the TRQ is subject to a zero or small duty.[14] "Above quota" imports (no longer applicable to Mexico under NAFTA) are subject to a prohibitive tariff (in FY2008, the equivalent of 86% on raw cane sugar and 76% on refined sugar), which effectively keeps out additional foreign sugar.

Sugar Market Developments

The policy debate in the next few months will revolve around how USDA should implement available authorities to either maintain the sugar import quota at already announced levels or increase quotas to increase available supplies. Because either alternative would affect market prices, the following lays out the current outlook for U.S. sugar demand, supply, and prices against historical experience.

Food Use

The United States is the fifth-largest user of sugar for food in the world, consuming an estimated 10.5 million ST in FY2009. While total sugar use declined during the first half of the last decade, in recent years deliveries to food/beverage manufacturers and consumers have trended upward (Figure 3). The increase in sugar demanded reflects population growth, food and beverage manufacturers' substitution of sugar for high-fructose corn syrup (HFCS) in response to changing consumer perceptions of HFCS, and consumers' shift towards diet drinks and bottled waters away from corn-sweetened soft drinks. As a result, the decline in HFCS use has been offset largely by an increase in total sugar use (Figure 3).

In calendar 2008, sugar alone accounted for about 41% of total U.S. sweetener consumption, up from 38% in 1999. This share includes the amount of sugar estimated to be present in net imports of sugar-containing products. HFCS and other corn sweeteners' share of domestic sweetener use stood at 40.5%, down from 46% a decade earlier.[15]

(FY2000-FY2010 projected)

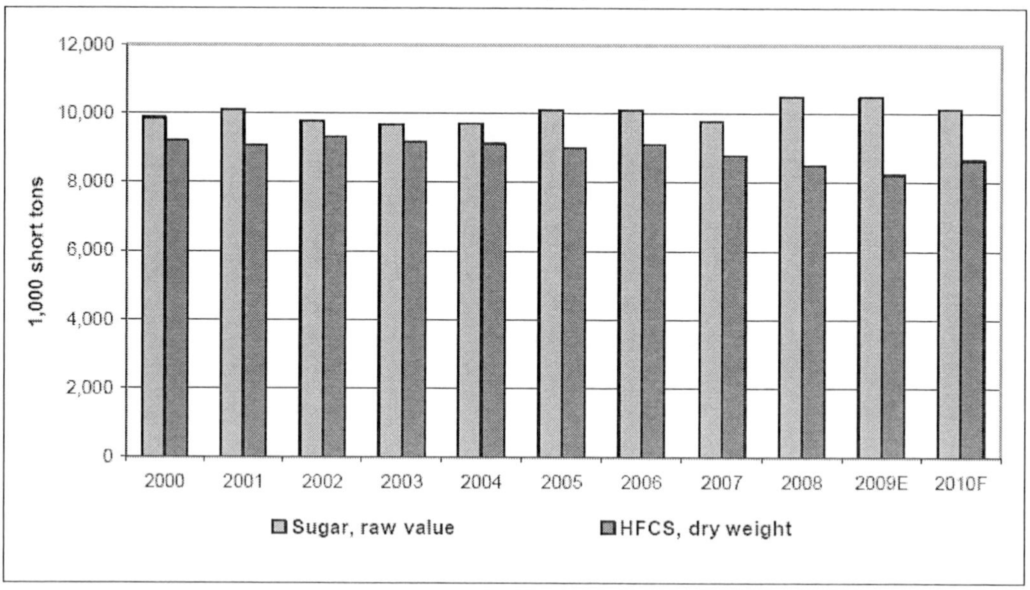

Source: USDA, World Agricultural Outlook Board, *World Agricultural Supply and Demand Estimates*, January 12, 2010, for recent year estimates on sugar deliveries; USDA, Economic Research Service, *Sugar and Sweeteners Outlook*, October 5, 2009, pp. 15 and 48, for HFCS deliveries for all years, and for sugar deliveries for earlier years adjusted by CRS for recent USDA definitional change.

Note: Sugar deliveries reflect January 2010 estimates; HFCS deliveries reflect mid-2009 estimates.

Figure 3. U.S. Deliveries of Sugar and HFCS for Food and Beverage Consumption.

Analysts differ over the FY2010 outlook for sugar food use. This key factor has implications for the policy decisions that USDA might make in the coming year. If available supplies appear to be tracking close to meeting USDA-projected U.S. food demand for sugar, pressure on USDA to increase imports may recede. However, if sugar use trends higher than now projected, USDA would face continued calls to allow for additional imports.

For FY2010, USDA projects food use of sugar will fall by almost 340,000 ST (3.25%), to 10.14 million ST. This outlook assumes there will be a reversal this year of what happened in the last two years (when direct consumption imports—primarily refined sugar—increased to substitute primarily for HFCS use, and as a result, boosted total sugar use). During FY2008 and FY2009, imports increased in response to USDA's August 2008 decision to temporarily allow additional shipments under the refined sugar TRQ, and as Mexican sugar mills significantly increased refined sugar shipments to the U.S. market, in large part to take advantage of higher U.S. prices. However, for FY2010 USDA projects refined sugar imports from Mexico for direct consumption will be smaller than in FY2009, and that entries under the raw and refined sugar TRQs will be capped by the minimum access levels laid out in U.S. WTO commitments (top of Table 1). As a result, USDA expects HFCS deliveries to increase to recapture the market share lost to sugar in the previous two years (Figure 3).

USDA acknowledges some uncertainty, however, about the extent to which HFCS use might increase during FY2010 at the expense of sugar. Consumer preferences may not "change sufficiently to accept HFCS in products that had [already] made the switch to sugar." USDA also notes that HFCS use will depend much on how far corn refiners go to "aggressively market their product to recapture [the market share] lost to sugar."[16] Reflecting this uncertainty, USDA admits that its FY2010 sugar use projection may be at the low end of a range.[17] Recent reports suggest that corn refiners have reduced contracted prices for some HFCS deliveries in 20 10,[18] a development that may make this corn sweetener more price competitive in certain end uses. The extent to which this may in fact reduce sugar use, however, will not be known for several months.

Another analysis projects that FY2010 food use will not fall as much as forecast by USDA. It estimates deliveries for food will fall by about 190,000 ST (1.8%), to about 10.3 million ST. While acknowledging that users will switch back to HFCS to some degree due to reduced imports of sugar from Mexico, this analysis suggests that HFCS use will continue to decline due to changing consumer preferences and in response to the doubling of the HFCS spot price since FY2005 (from 13 ¢/lb. to 26 ¢/lb. in FY2009).[19]

Supply

USDA projected during the early summer months of 2009 that U.S. sugar stocks relative to domestic food consumption would be much lower than usual at the end of both the FY2009 and FY2010 marketing years. This prompted calls by industrial sugar users for USDA to quickly allow for additional imports to head off sugar "shortages." Since there is a close correlation between ending stocks and the price of refined sugar in the last few months of a marketing year, their call for additional imports reflected more of a concern that domestic prices would quickly spiral much higher unless USDA took quick action.

During the marketing year, U.S. sugar stocks reach their lowest level in September. At this time, the sugar beet harvest begins in the Red River Valley straddling Minnesota and North Dakota, and in Michigan—the two main beet sugar producing regions in the country—followed by other beet- producing states. As factories process the harvested-and-then-stored sugar beets from late fall through early spring, the supply of refined sugar increases. Similarly, from November through early spring, sugarcane is cut and processed in Florida and

Louisiana (where 90% of U.S. raw cane sugar is produced), further adding to available sugar supply (Figure 4).

USDA estimates FY2009 ending stocks at 1.45 million ST, above the 1.33 million ST recorded at the close of FY2005 in the aftermath of Hurricane Katrina, which shut down two cane sugar refineries in New Orleans. For FY20 10, USDA projects ending stocks at 1.14 million ST, or 68% of the average ending stock level recorded in the six-year period (FY2003 to FY2008) covered by the 2002 farm bill (Figure 4).

However, Figure 4 presents only part of the story, in that absolute ending stocks do not capture the impact of the growth in domestic food/beverage demand for sugar on the supply available at the end of recent marketing years, and in turn, its impact on market prices. The "stocks-to-use ratio" takes both supply and demand into account, measuring the relationship between sugar stocks at the end of the marketing year to the total amount of sugar consumed during that year. Market participants view this ratio as a reasonably credible indicator of what the future price of sugar might be. In general, a higher ending stocks-to-use ratio suggests lower sugar prices in the future; conversely, a lower ratio implies higher prices later on.

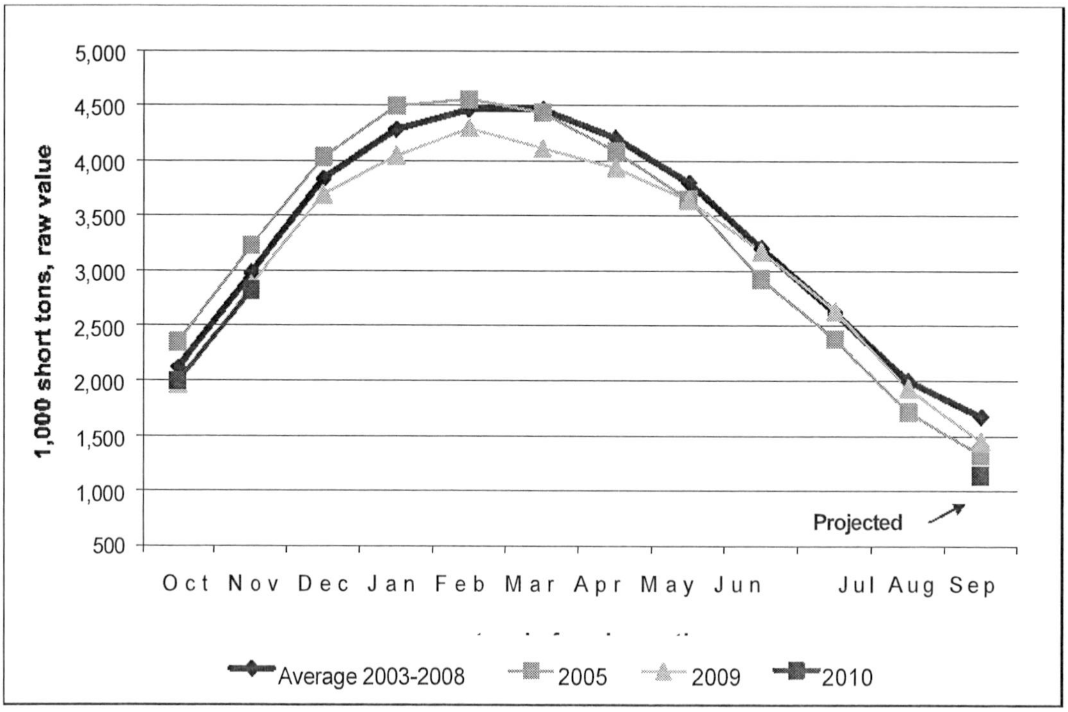

Source: USDA, Farm Service Agency, 'Sweetener Market Data' reports for FY2003 through November 2009; USDA, World Agricultural Outlook Board, *World Agricultural Supply and Demand Estimates*, January 12, 2010, for September 2010 projection.

Figure 4. Monthly U.S. Sugar Stocks (six-year average compared to selected years).

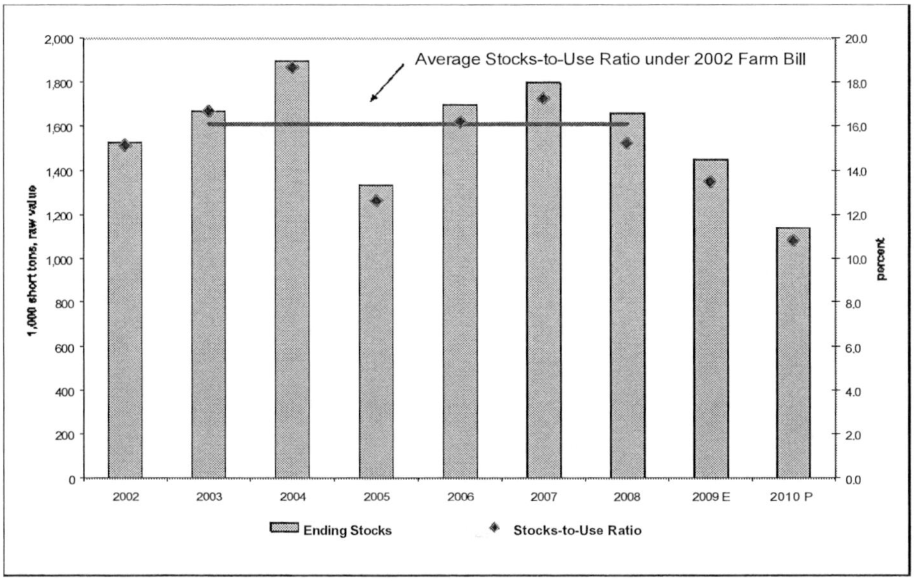

Source: USDA, Economic Research Service, for FY2002 through FY2008; USDA, World Agricultural Outlook Board, *World Agricultural Supply and Demand Estimates*, January 12, 2010, for FY2009 and FY20 10; average stocksto-use ratio derived by CRS.

Figure 5. U.S. Sugar Ending Stocks, and Stock-to-Use Ratios (sugar marketing years 2002-2010).

In implementing the sugar marketing allotment provisions of the last farm bill, USDA stated that its estimate of reasonable ending stocks will be set "at a level expected to preclude sugar loan collateral forfeitures"[20] (i.e., to ensure that sugar market prices were above effective price support levels so that the statutory directive to operate the sugar program at no cost to the U.S. Treasury was met). While USDA did not use an explicit stocks-to-use ratio to meet this objective, observers frequently noted that it was around 15%. During the six years covered by the 2002 farm bill, ending stock-to-use ratios ranged from a low of 12.6% in FY2005 to a high of 18.7% in FY2004. This ratio averaged 16.1% on an annual basis during this period (Figure 5).

To implement the 2008 farm bill provisions, USDA signaled that a sugar ending stocks-to-use ratio of about 14.5% is "the level [it] normally considers necessary to provide for a balanced domestic sugar market" and not likely to result in loan forfeitures.[21] For FY2009, lower beet production, the temporary loss of one cane refinery due to an explosion, and USDA's decision not to increase the import quota resulted in an ending stocks-to-use ratio of 13.5%. For FY2010, USDA currently projects an ending ratio of 10.8% (Figure 5), the lowest since FY1975.[22]

Price Developments

USDA estimates of tight domestic sugar supplies for year-end FY2009, and an even tighter supply outlook for FY20 10, have contributed to a run-up in the price of both raw cane sugar and refined sugar since July 2009. Prices began to rise in early August when a top official signaled that USDA would not "for now" increase the sugar import TRQ before

September 30, the end of the marketing year. Two subsequent developments pushed prices higher—the official's confirmation on August 23 that there will be no quota increase in FY2009, and USDA's announcement on September 25 that only the minimum quantities of sugar as spelled out in the U.S. TRQ commitment under the WTO will be allowed entry in FY2010.

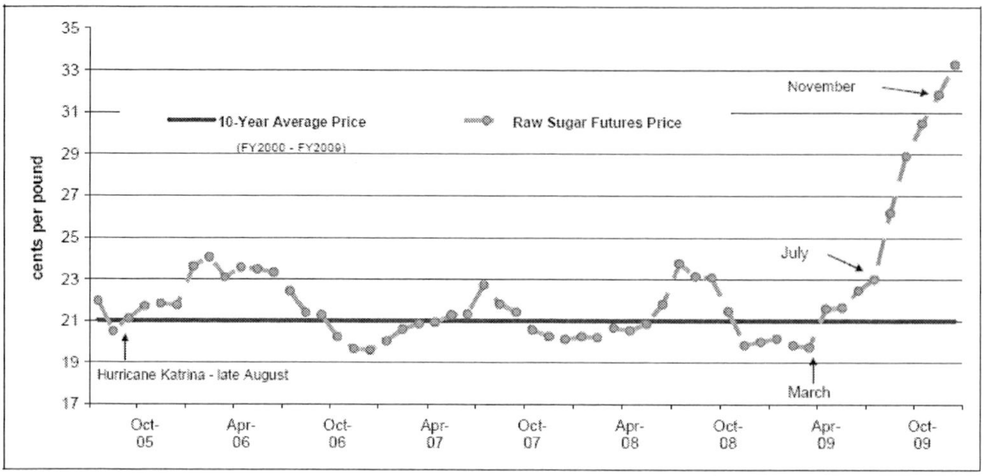

Source: USDA, Economic Research Service, "Sugar and Sweeteners Briefing Room;" 10-year average price derived by CRS.

Note: U.S. raw sugar price is the average monthly price for the nearby month #16 contract, traded in New York City on the IntercontinentalExchange (ICE).

Figure 6. U.S. Raw Cane Sugar Price Compared to 10-Year Average (monthly, July 2005 to December 2009).

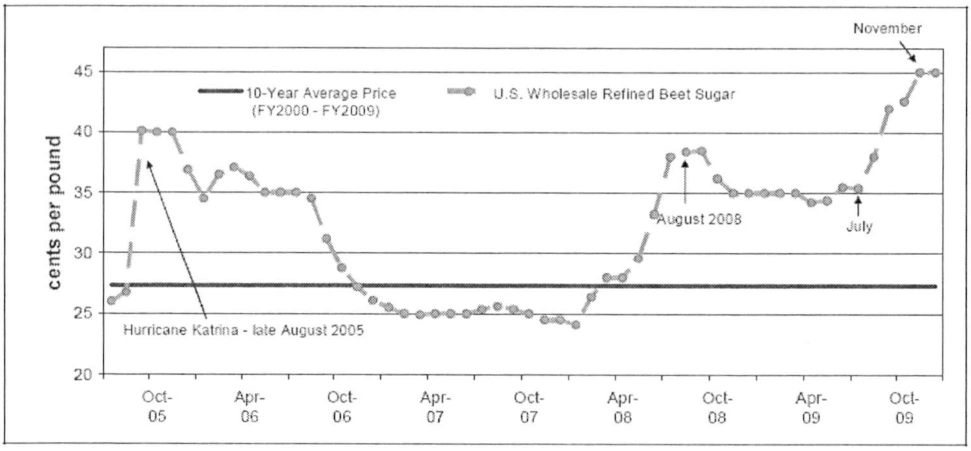

Source: USDA, Economic Research Service, "Sugar and Sweeteners Briefing Room;" 10-year average price derived by CRS.

Note: Simple average of the lower end of the range of quotations (Midwest markets) for all weeks in each month, as published by *Milling and Baking News*.

Figure 7. U.S. Refined Beet Sugar Price Compared to 10-Year Average (monthly, July 2005 to December 2009).

In response, from July to December 2009, the U.S. raw cane sugar price rose from a monthly average of 23.0 ¢/lb. to 33.3 ¢/lb., a 45% increase. As of mid-January 2010, the raw sugar futures price had climbed further to over 40.0 ¢/lb. To place in perspective, the 10-year (FY2000- FY2009) average price for raw cane sugar was 21.0 ¢/lb. (Figure 6). During the same six-month period, the U.S. wholesale refined beet sugar price for spot sales rose from a monthly average of 35.4 ¢/lb. to 45.0 ¢/lb., a 27% increase. In mid-January 2010, sales of refined sugar occurred at 48 ¢/lb. The last time refined sugar prices significantly increased was in September 2005, after Hurricane Katrina put two cane refineries in New Orleans out of commission for a few months. Then, refined sugar prices rose from almost 27 ¢/lb. (August 2005) to 40 ¢/lb. in September, where they remained for two months until beginning to decline in December 2005. In comparison, the 10-year (FY2000-FY2009) average wholesale price for refined beet sugar was 27.3 ¢/lb. (Figure 7).

Although U.S. sugar policy is structured to completely isolate the impact of lower world sugar prices on domestic sugar prices (by limiting imports), this can change whenever the world raw cane sugar price climbs above the U.S. effective support level. Drought caused by weak monsoon rains in India and slow harvesting caused by heavy rains in Brazil—both major players in world sugar trade—have reduced available global sugar supplies and propelled world sugar prices to a level not seen for almost three decades. As countries have scrambled to secure limited sugar stocks, the world raw cane sugar price (futures #11, traded in New York) has soared to the highest level in 29 years (i.e., since January 1981), reaching 29.0 ¢/lb on January 19, 2010. Since the United States must compete with other countries to cover its sugar import needs, U.S. cane refiners also face paying a premium to secure the foreign sugar that normally enters under quota. Though world raw sugar prices leveled off between August and November 2009, the continued climb in the U.S. raw cane sugar futures price (#16) since then indicates that a higher price level is needed to entice foreign exporters to ship sugar to the U.S. market against their quota allocations rather than to other destinations (Figure 8). This in turn has provided a price floor for U.S. refined sugar. As world raw sugar futures again began to climb in mid-December 2009, U.S. spot refined prices rose by another 3 ¢/lb, to 48 ¢/lb, in early January.[23] Analysts expect world sugar prices to remain at about current levels, if not rise slightly, until mid-year 2010, when Brazil is expected to begin to supply additional sugar as its cane harvest gets underway.

About one-quarter of the sugar used as food in the United States is purchased at the retail price by households in consumer-sized packages at supermarkets and grocery stores. Most of the remaining sugar is consumed as an ingredient in sugar-containing products sold by food and beverage manufacturing firms. These firms pay the wholesale price for sugar, which represents one cost component along with the price paid for each other ingredient used to produce a food product. Since the sugar content varies from product to product, an increase in the wholesale refined sugar price, if passed on by manufacturers, may mean that consumers pay more for some products at the retail level. However, price declines over the past year in several other agricultural commodities purchased by food manufacturers and used with sugar in baked goods, for example, may have offset increased sugar costs, and resulted in little change in the retail price of some products. In recent years, the retail price of refined sugar has lagged behind changes in the wholesale price, and since early 2008, it has increased at a slower rate than the rise in the wholesale price (Figure 9). During FY2009, the retail price averaged 56.0 ¢/lb., compared to the 10-year average (FY2000-FY2009) of 46.5 ¢/lb.

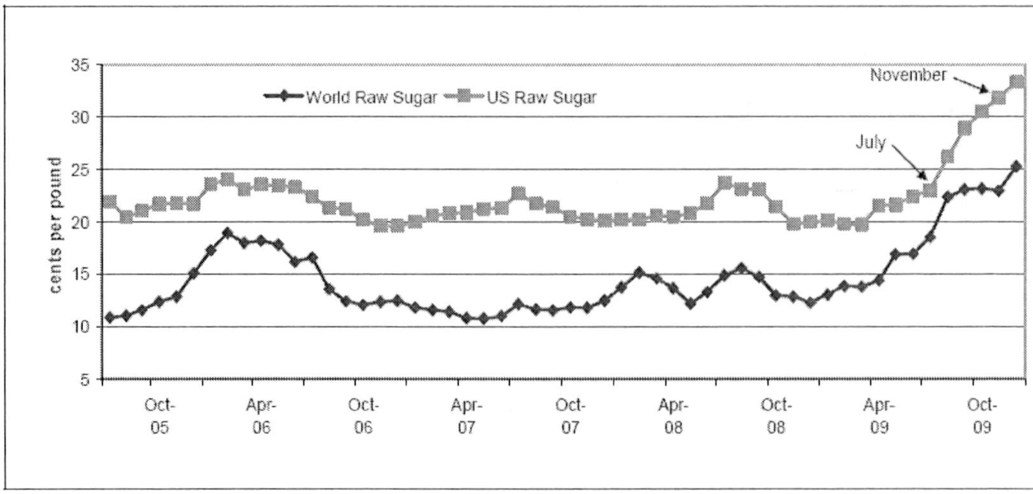

Source: USDA, Economic Research Service, "Sugar and Sweeteners Briefing Room."

Note: Prices shown are for futures #11 (world) and spot #16 (U.S.), traded in New York City on the IntercontinentalExchange (ICE).

Figure 8. World and U.S. Raw Sugar Prices (monthly, July 2005 to December 2009).

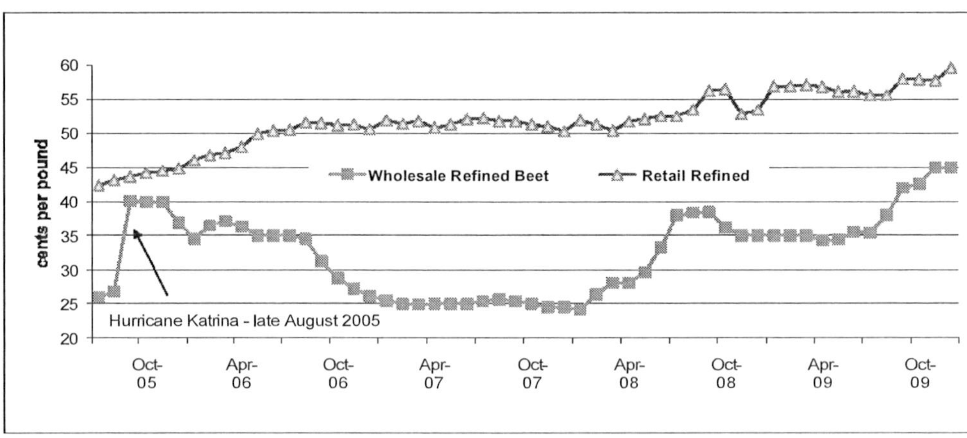

Source: USDA, Economic Research Service, "Sugar and Sweeteners Briefing Room;" Department of Labor, Bureau of Labor Statistics, for recent months' retail prices.

Note: Retail refined price is for *all* refined sugar, irrespective of whether processed from sugar beets or raw cane sugar.

Figure 9. U.S. Wholesale and Retail Refined Sugar Prices (monthly, July 2005 to December 2009).

Price Outlook

Given the prospect of continued high world prices for the next several months and a projected tight domestic supply outlook, U.S. raw and refined sugar prices are expected to remain high by historical standards through FY20 10. The likelihood of higher sugar prices is signaled in large part by prices quoted for world and domestic raw sugar futures contracts.

World raw sugar (futures #11, ICE, for the March, May, and July 2010 contracts) recently traded between 29 ¢/lb. and 24 ¢/lb., driven by the global outlook for tight supplies for several months out. Domestic raw sugar (futures #16, ICE, for the March, May, and July 2010 contracts) has in recent weeks traded between 41 ¢/lb. and 34 ¢/lb., supported largely by the run-up in the world raw sugar futures price and speculative trading.

While there is no futures market for refined sugar in the United States, pricing agreed upon between sugar processors and industrial sugar users for near-term sales of refined beet sugar serves as a reasonable proxy for what refined prices might be a few months out, barring any noticeable change in the domestic supply picture.[24] Recent spot sales of refined sugar were made at 48 ¢/lb. with signs that some processors would not have sugar to sell beyond the first quarter of 2010.[25] However, any change to current policy (see "2008 Farm Bill Sugar TRQ Provision") and in projected imports from Mexico (see "Sugar Imports from Mexico") could alter the outlook for domestic raw cane and refined beet sugar prices.

Price Setting

Common practice is for sugar processors and industrial sugar users to negotiate contracts that detail the sales price for specific quantities of refined sugar to be delivered at specified future points in time. Because these contracts frequently are negotiated far in advance (covering shipments up to more than a year into the future), sellers and buyers weigh what the following factors portend for future price levels. Before agreeing on price, both sides take into account projections of future U.S. sugar supply and demand, particularly ending stock levels; expectations about what USDA might decide with respect to sugar import quota levels; implications of future raw cane sugar prices as signaled by price quotes in the futures market; the nature of existing commercial relationships; and market intelligence on the quantity of refined sugar expected to enter from Mexico. During these negotiations, each sugar processor seeks to sell much of the sugar it is permitted to under its marketing allotment to minimize what is held in inventory at the end of the year, at the highest possible price. Similarly, each user aims to line up guaranteed deliveries of sugar to cover its needs over time, at the lowest price possible. Anecdotal reports suggest that a large portion of the sugar delivered each year is marketed on this basis.

For example, from 75% to nearly all of the expected needs of industrial sugar users apparently have been booked for calendar 2010. Retail buyers (i.e., wholesale grocers, retail grocers, and chain stores) reportedly have less of their sugar needs covered for 2010, which means that "a significant amount of total demand" is yet to be filled. However, some buyers indicated that "they planned to delay large-scale [purchases] until after April 1, 2010," when USDA can allow for additional sugar imports, in the hope that a decision "might push prices lower"[26] (see "2008 Farm Bill Sugar TRQ Provision"). However, some users, such as smaller food/beverage firms and retailers, prefer to cover some or most of their sugar requirements on an as-needed basis. This means they pay the current ("spot") price for refined sugar, which might be higher or lower than the price negotiated by other users under the longer-term contractual arrangements described above.

Beyond this year, sugar processors and users reportedly continue to negotiate "significant" bookings of refined sugar for delivery in 2011 at 35 ¢/lb. or higher. Some processors are urging buyers to now cover their needs into 2011, mentioning the volatile world sugar supply situation and higher prices.[27]

ISSUES

Sugar users continue to press their case for USDA to increase sugar imports, arguing that quick action is needed to restore supplies to more normal levels. Sugar processors maintain that the domestic market is adequately supplied now that the harvesting and processing of the 2009 crops are underway. Although these contrasting views focus more on the availability of sugar in the current year, the underlying issue driving the debate is more about the level of the price of sugar. Sugar processors seek to maintain their advantage in negotiating sales prices with users. Food and beverage firms want prices for refined sugar that are lower and closer to the historical average. Members of Congress have weighed in with letters to USDA in support of each side's position.

U.S. sweetener trade with Mexico—now free in both directions—continues to be watched carefully because of its impact on U.S. sugar supplies and prices. The uncertainty that USDA faces in projecting how much sugar from Mexico might enter the U.S. market has implications for how it manages U.S. sugar policy.

2008 Farm Bill Sugar TRQ Provision

One new farm bill provision prescribes USDA's authority in administering the sugar TRQs established to meet U.S. WTO trade commitments.[28] It requires USDA to set at the beginning of each marketing year (October 1) the quotas for raw cane sugar and refined (excluding specialty) sugar at the minimum WTO levels (top of Table 1). It also lays out the steps to be followed to increase imports in case there is a sugar shortage. The details differ, depending upon when in the year the Secretary of Agriculture exercises this authority.

In the event of an "emergency shortage of sugar" caused by a war, flood, hurricane, or other natural disaster, or other similar event as determined by the Secretary, during the first half of the marketing year (October 1 to March 31), USDA must take action to increase the supply of sugar. One action that can be taken is to increase the raw cane sugar TRQ to compensate for insufficient domestic cane and beet sugar output. If this step still leaves a sugar shortage, and the sale of domestic sugar "has been maximized" and U.S. raw cane sugar refining capacity also "has been maximized," the Secretary may increase the refined sugar TRQ.

In the second half of the marketing year (April 1 to September 30), if a sugar shortage still exists and the marketing of domestic sugar has been maximized, the Secretary can only increase the raw cane sugar TRQ. USDA is required to ensure that any increase in imports under this authority "will not threaten to result in" a fall in domestic sugar prices below loan forfeiture levels (i.e., when sugar processors can exercise their right to hand over sugar pledged as collateral for price support loans to the U.S. government as full payment).

In issuing the sugar program regulations to implement the 2008 farm bill changes,[29] USDA did not define this provision's key terms or any details indicating the circumstances under which an increase in sugar imports under this policy change would be implemented.

This enacted provision removed considerable discretionary authority that USDA previously exercised in making sugar TRQ decisions on import quantities and timing of entry by sugar type (raw cane versus refined). As now structured, USDA is directed to ensure that most imports enter in the form of raw cane sugar rather than refined sugar. The sugar industry sought this provision to reduce the considerable uncertainty associated with how much sugar actually enters from Mexico under NAFTA, and what portion enters as refined sugar. Since most cane refineries are now a key part of vertically integrated firms owned by raw sugar processors and/or sugarcane producers, the 2008 farm bill's TRQ directive also ensures that cane refineries (which process raw sugar into refined sugar) in an emergency situation can receive enough raw sugar in order to more fully utilize their operating capacity. Further, limiting the entry of refined sugar only when there is an emergency sugar shortage enhances the competitive position of domestic beet sugar refiners (and by extension, cane refiners) in securing the best possible price from purchasers.

Sugar Users Call for Immediate Import Increase

Sugar users, beginning in March 2009, have made repeated requests for USDA to increase import quotas for raw and refined sugar in order to head off a shortage in the domestic sugar supply. Pointing out that USDA projections showed historically low ending-stock levels for FY2009 and an even lower level for FY20 10, the Sweetener Users Association (SUA) urged that quick action be taken to head off the potential for market disruptions, particularly should a hurricane occur. It argued this was necessary because of the lead time required to purchase, ship, and refine raw sugar before refined sugar actually becomes available to users.

In late October 2009, the SUA called upon USDA to "responsibly manage the sugar program by announcing a significant increase in raw and refined sugar import quotas in coming weeks." It argued that even with the 2008 farm bill TRQ provision, the Secretary of Agriculture "still has ample and clear cut authority to increase TRQs at any time" by how he defines "emergency" and by using a sugar tariff headnote that gives him authority to increase imports if he concludes the U.S. sugar supply "may be inadequate to meet domestic demand at reasonable prices" (see footnote 11). Noting that the highest refined sugar prices in over 25 years do not represent "'reasonable prices' under any normal understanding of that term," the SUA stated that the headnote's "condition for a TRQ increase would appear clearly to be met." Further, it contended that "there would be ample scope within the farm bill language itself to consider" the outlook of record-low FY2010 ending stocks as constituting an "emergency shortage" and sufficient to announce an import quota increase. Referring to the stocks-to-use ratio as the best guide to follow to determine how much to increase the quota, the SUA argues current market conditions warrant a 1.0 to 1.2 million ST increase in the raw and refined sugar TRQs (for imports from countries eligible to export against their shares of the WTO import quota commitment). Its report concludes that "[f]ailure to act in a timely manner risks locking in high sugar costs for consumers and for food and beverage manufacturers for another year."[30]

Sugar Producers View Supplies as Adequate

Just before the current marketing year began, producers and processors maintained that there is plenty of sugar for sale and that with harvests underway, the domestic market "will remain adequately supplied for the foreseeable future." Representing them, the American Sugar Alliance (ASA) noted that "USDA projects a sugar surplus again this crop year." It applauded the decision to set sugar import quotas at their minimum levels, stating that this follows the congressional intent of the TRQ provision included in the 2008 farm bill. ASA again noted that this authority serves the purpose of taking "the guesswork out of TRQ decisions" and ensures that USDA makes them "with the best market information available."[31] Since the supply and demand outlook is more certain by the middle of the marketing year, ASA advocates that USDA wait until April 1, 2010, or even later, before making any decision on whether to allow increased imports.

Implications of the Timing of a TRQ Decision

When USDA set the FY2010 raw and refined sugar import quotas at the minimum WTO-commitment levels, it acknowledged that the U.S. market "will require additional supplies of sugar." Accordingly, it plans to "closely monitor stocks, consumption, imports, and all sugar market and program variables on an ongoing basis."[32] Of these factors, the uncertainty surrounding how much sugar could be exported to the U.S. market by Mexican mills is the most significant (see "Sugar Imports from Mexico").

Reports have surfaced that USDA may announce a quota increase in January 2010, but others think that decision will wait until April. In the four months since USDA announced the FY2010 sugar TRQs, USDA has raised its supply projection for the year by 2.2%. Lower beet sugar output due to harvesting problems in freezing conditions and lower sugar content is expected to be more than offset by increased raw cane sugar output in Louisiana and additional imports from Mexico. Changes in supply, and possibly in demand, can be expected. At present, USDA still projects extremely low ending stocks compared to historical experience (Figure 5).

A USDA decision on whether or not to increase the sugar TRQ, and then its timing, will continue to be a significant factor influencing negotiations between sugar processors and industrial sugar users on the sales price for product to be delivered over the next 12- to 24-month period. If USDA does plan to eventually increase the import quota, a delay in making a decision (because of the uncertainty raised) strengthens the hand of processors as prices are negotiated for sugar that users seek to secure in the short term and beyond. Expecting that USDA will not make a decision until April 2010, some sellers have raised price offers or contracted sales at higher prices from users seeking sugar to cover immediate needs. To illustrate, spot prices for the near-term delivery of refined sugar stood at 48 ¢/lb. in mid-

January 2010, compared to 35 ¢/lb. one year earlier. However, most buyers reportedly are reluctant to contract on a longer-term basis .[33]

Further, if and when USDA announces a TRQ increase, the impact of the additional imports on increased supply in subsequent months would depend on the size of the increase. The focus will be on how much the announced increase covers the difference between currently projected ending stocks and what supply would be needed to conclude FY2010 with a 14.5% stocks-to-use ratio— the informal benchmark used by USDA. At present, this difference is about 400,000 ST (assuming there is no additional increase in imports from Mexico beyond that already projected and no change in USDA's food use estimate). With additional imports, prices could fall some, but the extent of the drop would depend on the amount that actually enters against the announced quota and how quickly sugar is imported and refined. However, it is quite possible that with world sugar prices at historically high levels because of tight supplies, the additional demand created by a U.S. quota announcement might mean cane refiners would have to pay more for their primary input. In turn, refiners likely would try to pass on that additional cost to industrial users and retail buyers.

Sugar Imports from Mexico

Free Trade under NAFTA

Unrestricted imports of sugar from Mexico since January 1, 2008, have introduced considerable uncertainty into USDA's efforts to manage the sugar program to ensure that market prices stay above effective support levels (i.e., that sugar processors have no incentive to forfeit on their loans if prices are lower). Since then, Mexico's share of U.S. sugar imports has fluctuated widely—from 34% in FY2008 to an estimated 51% in FY2009 (Figure 10). While sugar imports under U.S. trade commitments to the WTO and FTA trading partners have been stable and predictable over the last few years, estimates of entries from Mexico can change substantially during a year's time. For example, as market conditions changed during FY2009, USDA increased its estimate of imports from Mexico by 21/2 times, from an initial 550,000 ST projection early in the year to 1.4 million ST at the end.

Record sugar imports from Mexico in FY2009 were prompted by two major factors. Faced with the need to generate cash to meet financial obligations to cane growers and suppliers, Mexican sugar mills resorted to selling sugar in the U.S. market, even at times when U.S. prices were lower than in Mexico. More important, from late spring 2008 through early summer 2009, Mexican mills took advantage of the higher U.S. refined sugar price to significantly increase sales of refined sugar to the U.S. market. As the price in both markets has converged since late summer, Mexican sales to the United States have declined (Figure 11). As of late December 2009, traders stated there was no sign of Mexican sugar shipments to the U.S. market, in part due to more sugar from the new cane harvest being kept to rebuild depleted stocks. One report notes that while Mexican sugar prices have fallen as more new sugar enters the marketplace, there "remained little price advantage to ship sugar north."[34]

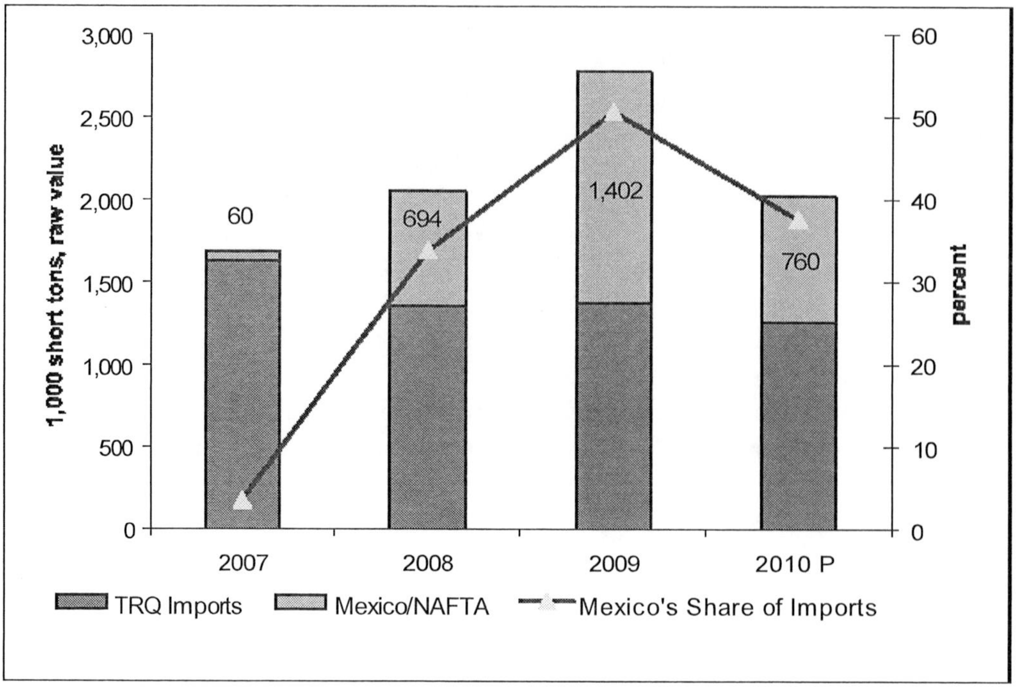

Source: USDA, Economic Research Service, "Sugar and Sweeteners Briefing Room" for FY2007; USDA, World Agricultural Outlook Board, *World Agricultural Supply and Demand Estimates*, January 12, 2010, for more recent year data; Mexico's share is derived by CRS.

Note: TRQ imports reflect actual entries based upon WTO and DR-CAFTA trade commitments.

Figure 10. Mexico's Share of U.S. Sugar Imports for Food Has Fluctuated (FY2007 to FY2010. projected).

For FY2010, USDA projects 760,000 ST in sugar imports from Mexico, slightly more than half the previous year's level. This represents 37% of expected total U.S. sugar imports for food use (Figure 10). While still a sizable amount, the expected lower import level is based on the view that the significant drawdown in Mexican sugar stocks during FY2009 (with a large portion having been exported to the U.S. market), even with projected average sugar output in FY2010, means that Mexico will have less available to export to the U.S. market. Although imports authorized by the Mexican government covered the country's immediate needs through year-end 2009 until new-crop sugar became available, USDA projects that Mexican stocks at the end of FY2010 will still be much below half of recent historical experience. This suggests that Mexican sugar prices will remain higher than in recent years, a factor that could affect how much sugar actually flows north. If Mexican prices stay above U.S. price levels, sellers can be expected to meet the demand found in the more profitable Mexican sugar market. Conversely, if U.S. prices rise above Mexican levels, Mexican mills have an incentive to ship sugar to the United States.

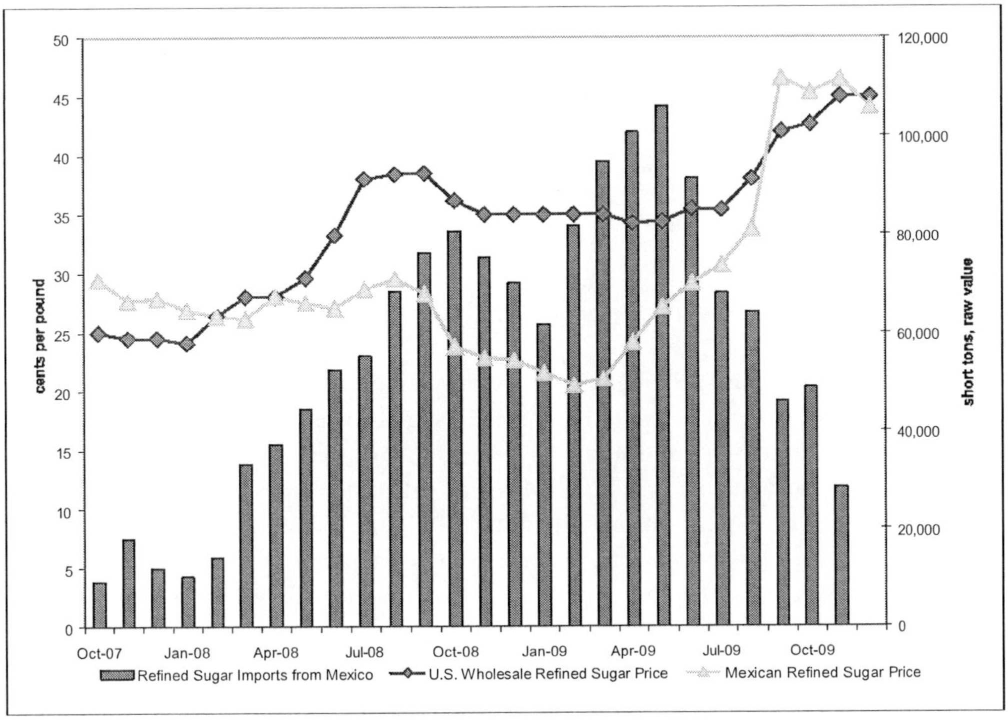

Source: USDA, Foreign Agricultural Service, for monthly import data; USDA, Economic Research Service, for U.S. and Mexican prices.

Notes: Imports of *refinado* (refined sugar) from Mexico are estimates based upon mode of entry (containers, railcars, or trucks)—likely to end-users rather than to U.S. cane refineries—as recorded by U.S. Customs and Border Protection, and converted from metric to short tons. U.S. wholesale price is for refined beet sugar, Midwest markets, as published by *Milling and Baking News*. Mexican wholesale price is for bulk quantities of *refinado* quoted in Mexico City and reported by "Servicio Nacional de Informacion de Mercados (SNIIMECONOMICA)," and converted from pesos to U.S. dollars by ERS using exchange rates published by the Board of Governors of the Federal Reserve System.

Figure 11. U.S. Imports of Refined Sugar from Mexico Have Increased When U.S. Price is Higher than Mexican Price (FY2008 to December 2009).

Three factors could alter this outlook. They introduce elements of uncertainty into USDA's projection on sugar imports from Mexico this year. First, irrespective of the relationship between Mexican and U.S. sugar prices, some Mexican mills still may ship sugar north to maintain the commercial trading relationships they have already developed and to generate dollars needed to cover their financial obligations. This may mean more imports from Mexico than USDA currently projects. Second, U.S. exports of high-fructose corn syrup (HFCS)—a competitive sweetener—to Mexico, if priced lower than Mexican-produced sugar, could displace some of the Mexican- produced sugar now used by that country's soft-drink manufacturing sector. This substitution occurred in the United States in the early 1980s, when cheaper HFCS almost completely substituted for the more expensive sugar that the U.S. soft drink sector had historically used. With the significant rise in the Mexican sugar price during 2009, and reported weak HFCS sales and prices in the United States, demand for HFCS appears to be growing in Mexico. This in turn reportedly is expected to result in increased HFCS exports to the Mexican market in coming months.[35] To the extent that HFCS displaces sugar in Mexico's soft drink sector, any resulting surplus of Mexican sugar would have an incentive to move north to the U.S. market. Third, the Mexican government may take steps to rebuild domestic sugar stocks (e.g., license exports to the U.S. market, or announce another import quota). Each would have differing implications for the amount of Mexican sugar that may flow north.

"Substitution" and Governments' Coordination of Sugar Policies

U.S. policymakers and U.S. sugar industry officials claim that Mexico is "substituting" lower- priced imported sugar to cover its domestic sugar needs. They point out that over the last year, Mexican mills exported much more sugar than expected to the U.S. market—a development that contributed to a fall in Mexican sugar stocks to below normal levels. In turn, the Mexican government substantially increased imports in order to avert shortages and to reduce the prices paid by users and consumers. Some Members of Congress have called for this "loophole" in NAFTA to be addressed. [36] This practice, they contend, makes it very difficult for USDA to manage sugar policy, which in turn affects projections of U.S. sugar supply and the outlook for U.S. prices.

These statements were made against the backdrop of the Mexican government's sudden recognition in early August 2009 that there would not be sufficient stocks of sugar on hand to cover the country's needs before sugar from the next sugarcane crop becomes available in December 2009.[37] To respond, the government announced three sugar import quotas through mid- September to allow for the entry of 900,000 MT by year-end 2009. However, the auction for the last announced quota was cancelled in late September, reportedly to address the Mexican sugar production sector's concern that any additional imports would lower the price received this year and next. As a result, only 550,000 MT will be allowed to enter under these temporary quotas to rebuild very low stock levels.

The "loophole" refers to a provision negotiated in the more recent FTAs (e.g., with Chile, Morocco, the DR-CAFTA countries, and Peru) but not included in the NAFTA text negotiated more than 15 years ago with Mexico. The newer provision limits the preferential (i.e., duty-free and quota) access each FTA partner has to the U.S. market only to its net sugar trade surplus. This surplus is each country's volume of exports to the world, minus the volume of its imports from the world, of sugar and specified sugar-containing products, with adjustments for its sweetener trade with the United States, in the most recent year. If exports are greater than imports, the FTA partner's access to the U.S. market is limited to this surplus or to the U.S. quota commitment set in the FTA for the year, whichever is lower.

Acknowledging that formal efforts to address the "loophole" would reopen the controversial issue of renegotiating NAFTA, parts of the U.S. and Mexican sugar production sectors recently presented to their respective governments a detailed framework for coordinating both countries' sugar policies to address the "substitution" issue. The most significant recommendation would have both governments cooperate when they determine (1) that a sugar shortage exists in the combined U.S.-Mexican sugar markets, and (2) the amount of "non-NAFTA" imports required to eliminate that shortage. Another would have both countries modify their sugar re-export programs so that sugar-containing products manufactured with lower-cost world (i.e., "non-NAFTA") sugar in the United States and Mexico are exported only to outside the two-country region. Other recommendations call for both governments to consult at least every three months to review data on both countries' sugar markets; to establish a permanent joint sugar commission to coordinate national sugar policies, monitor implementation of the framework's objectives, and handle any disputes that may arise under the framework; to consult about the likely impact of future trade agreements on how their sugar programs work; and to work together to improve the collection and publication of reliable data on each country's sugar supply and use, including mandatory and enforceable reporting requirements for sugar producers.[38]

U.S. sugar producers state that these recommendations can be put into place without making any changes to NAFTA, do not require any legislation, and maintain "an open NAFTA market" by not restricting the flow of sugar between Mexico and the United States. They also note that this proposal was structured so as not to affect other U.S. commodities that view Mexico as a major market, including corn sweeteners. U.S. sugar users, however, argue that this proposal would change how NAFTA operates, since both governments "would not be able to independently operate their sugar programs." Further, they characterize the recommendations as interfering with Mexico's right under NAFTA to meet its needs with unlimited imports and for its sugar mills to ship all of their available sugar to the U.S. market.[39]

End Notes

[1] For background, see CRS Report RL34 103, *Sugar Policy and the 2008 Farm Bill*, by Remy Jurenas.

[2] This is a type of loan where processors pledge a commodity as collateral to obtain a loan from USDA's Commodity Credit Corporation (CCC) at a commodity-specific, per-unit loan rate. The borrower may repay the loan, with interest, within a specified period and regain control of the commodity. Alternatively, the commodity can be forfeited to the CCC with no penalty if market prices fall below support levels at the end of the term. The U.S. government takes no recourse beyond accepting the commodity as full settlement of the loan.

[3] USDA, Farm Service Agency, "USDA Announces 2009 - Crop Sugar Loan Rates", September 28, 2009, available at http://www.fsa.usda.gov/FSA/printapp?fileName=nr_20090928_rel_0473.html&newsType=newsrel.

[4] Imperial Sugar's facility, located in Savannah, Georgia, supplies almost 10% of the refined sugar consumed in the United States. The company expects its bulk sugar manufacturing line to be fully operational in the first quarter of calendar year 2010.

[5] For explanation, see "Import Quotas" in CRS Report RL34 103, *Sugar Policy and the 2008 Farm Bill*, by Remy Jurenas.

[6] One short ton equals 2,000 pounds. Raw value refers to the conversion of refined sugar into an equivalent that can be used alongside raw cane sugar.

[7] *Federal Register*, "Commodity Credit Corporation - Domestic Sugar Program—2008-Crop Cane Sugar and Beet Sugar Marketing Allotments and Company Allocations," September 10, 2009, available at http://edocket.access.gpo.gov/2009/pdf/E9-21761.pdf.

[8] USDA, "USDA Announces Initial Fiscal Year 2010 Sugar Marketing Allocations", September 10, 2009, available at http://www.fsa.usda.gov/FSA/newsReleases?area=newsroom&subject=landing&topic=ner&newstype= newsrel&type= detail&item=nr_20090928_rel_0477.html.

[9] USDA, Foreign Agricultural Service, "USDA Established Fiscal Year 2010 Tariff-Rate Quotas for Raw, Refined, and Specialty Sugar and the Domestic Sugar Overall Allotment Quantity", September 25, 2009, available at http://www.fas.usda.gov/scriptsw/PressRelease/pressrel_dout.asp?Entry=valid&PrNum=0244-09.

[10] A TRQ combines two policy instruments used to restrict imports: quotas and tariffs. The quota component works together with a specified tariff level to provide the desired degree of import protection. Imports entering under the quota portion of a TRQ are usually subject to a lower, or sometimes a zero, tariff rate. This "in-quota" amount represents the minimum that a country has committed to allow to enter under multilateral or other trade agreements. Imports above the quota's quantitative threshold (referred to as "above-quota") face a much higher (usually prohibitive) tariff.

[11] Harmonized Tariff Schedule of the United States (HTS), Chapter 17, Sugars and Sugar Confectionery, Additional U.S. Note 5, p. 17-2, available at http://www.usitc.gov/publications/docs/tata/hts/bychapter/0911C17.pdf.

[12] USTR, "USTR Announces FY 2010 Tariff-Rate Quota Allocations for Raw Cane Sugar, Refined and Specialty Sugar and Sugar-Containing Products", September 25, 2009, available at http://www.ustr.gov/about-us/press-office/ press-releases/2009/september/ustr-announces-fy-20 1 0-tariff-rate-quota-allocat.

[13] Congress has not yet considered the FTAs with Panama and Colombia, which, if approved, would grant additional access for each country's sugar into the U.S. market.

[14] A most favored nation tariff of 0.66¢ per pound is imposed *only* on imports of raw cane sugar from three countries that ship against their allocation of the WTO raw cane sugar TRQ (Argentina, Australia, and Brazil).

[15] Derived by CRS from USDA, ERS, *Sugar and Sweeteners Outlook*, "U.S. Sweetener Demand," June 2, 2009, p. 14, as accessed at http://www.ers.usda.gov/Publications/SSS/2009/SSS255.pdf. Other corn sweeteners consumed are dextrose and glucose syrup. Other caloric sweeteners are pure honey and edible syrups. High-intensity artificial sweeteners accounted for almost 18% of all sweetener use, up from just above a 15% share a decade earlier.

[16] In other words, if corn refiners price their HFCS at a substantial discount to refined sugar, they could regain some of their market share.

[17] USDA, Economic Research Service, *Sugar and Sweeteners Outlook*, SSS-256, October 5, 2009, pp. 12-19.

[18] *Milling and Baking News*, "Ingredient Week—Sweeteners," January 12, 2010, p. 46.

[19] Promar International, *North American Sweetener Report*, September 2009, pp. 2-3, 8, November 2009, pp. 4-5, December 2009, p. 2. This firm provides consulting services to food and agribusiness firms, including those that use sugar and other sweeteners.

[20] *Federal Register*, "Commodity Credit Corporation, 2002 Farm Security and Rural Investment Act of 2002 Sugar Programs and Farm Facility Storage Loan Program," August 26, 2002, p. 54927.

[21] *Federal Register*, p. 15359 (see footnote 29 for link). CRS derived this ratio using the parameters that the Farm Service Agency spelled out in the "Background" section published in this final rule.

[22] USDA, World Agricultural Outlook Board, *World Agricultural Supply and Demand Estimates (WASDE)*, November 10, 2009, p. 16, accessed at http://usda.mannlib.cornell.edu/usda/current/wasde/wasde-12-10-2009.pdf; USDA, ERS, *U.S. Sugar Statistical Compendium*, August 1991, p. 117.

[23] *Milling and Baking News*, "Ingredient Week—Sweeteners," November 3, 2009, p. 46, November 10, 2009, p. 38; January 12, 2010, p. 46.

[24] Also, the refined cane sugar price closely tracks that of refined beet sugar, since they are identical products and compete against each other in many U.S. regional markets.

[25] "Ingredient Markets—Sweeteners" columns in *Milling and Baking News*, January 12, 2010, p. 46, and in *Food Business News*, January 19, 2010, p. 50.

[26] *Milling and Baking News*, "Ingredient Week—Sweeteners," November 10, 2009, p. 38, November 24, 2009, p. 46; December 1, 2009, p. 42; *Food Business News*, "Ingredient Week—Sweeteners," December 8, 2009, p. 50.

[27] *Milling and Baking News*, "Ingredient Week - Sweeteners," September 29, 2009, p. 30, November 17, 2009, p. 42; December 22, 2009, p. 46.

[28] Section 1403(j) of P.L. 110-246, the 2008 farm bill; 7 U.S.C. 1359kk.

[29] *Federal Register*, "Sugar Program – Final Rule," issued by USDA's Commodity Credit Corporation, April 6, 2009, available at http://edocket.access.gpo.gov/2009/pdf/E9-7633.pdf.

[30] Promar International, "Responsible Management of the Sugar Program Requires A Quota Increase," October 28, 2009, available at http://www.sweetenerusers.org/ Responsible%20Management%20of%20the%20Sugar%20Program%20(3).pdf.

[31] American Sugar Alliance, "ASA Statement on FY2010 TRQ Announcement," September 25, 2009, accessed at http://www.sugaralliance.org/newsroom/asa-statement-on-fy2010-trq-announcement.html.

[32] USDA, Foreign Agricultural Service, "USDA Establishes Fiscal Year 2010 Tariff-Rate Quotas for Raw, Refined and Specialty Sugar and the Domestic Sugar Overall Allotment Quantity," September 25, 2009, available at http://www.fas.usda.gov/scriptsw/PressRelease/pressrel_dout.asp?Entry=valid&PrNum=0244-09.

[33] "Ingredient Week—Sweeteners" columns in *Milling & Baking News*, November 17, 2009, p. 42; December 8, 2009, p. 50; January 12, 2010, p. 46.

[34] *Food Business News*, "Ingredient Markets—Sweeteners," January 5, 2010, p. 46.

[35] *Food Business News*, "Ingredient Week—Sweeteners," December 22, 2009, p. 46.

[36] *Inside U.S. Trade*, "USDA: NAFTA Sugar 'Loophole' Complicates Estimates of Mexican Imports," August 7, 2009; American Sugar Alliance (ASA), "Sen. Crapo: Mexican Sugar Issues Must be Addressed," August 4, 2009, accessed at http://www.sugaralliance.org/newsroom/sen-crapo-mexican-sugar-issues Sugaronline.com, "US: Beet growers want NAFTA loop-hole closed," October 15, 2009; ASA, The Sugar Beat, "U.S. and Mexican Sugar Industries Agree that NAFTA Can Work Better," November 2009, accessed at http://www.sugaralliance.org/thesugar-beat/us-and-mexican-sugar-industries-agree-that-nafta-can-work-better.html.

[37] Mexico's sugar output, though trending upward, does vary from year to year, depending upon weather and growing conditions. Mexican government policy also is to hold three months worth of sugar stocks in reserve

and to allow sugar imports when needed to meet demand and lower prices. Also, U.S. sugar processors are now free to export sugar to Mexico to take advantage of the occasional higher prices there.

[38] Bureau of National Affairs, *International Trade Daily*, "U.S. and Mexican Sugar Industries Develop Joint NAFTA Recommendations," November 5, 2009.

[39] Sugaronline.com, "US and Mexican trade agree on industry overhaul," November 3, 2009; *Inside U.S. Trade*, "U.S., Mexican Sugar Industries Seek Closer Government Coordination," November 6, 2009; ASA, The Sugar Beat, "U.S. and Mexican Sugar Industries Agree that NAFTA Can Work Better," November 2009, accessed at http://www.sugaralliance.org/the-sugar-beat/us-and-mexican-sugar-industries-agree-that-nafta-can-work-better.html; Sweetener Users Association, "Broad-Based Public Interest, Business, Food Industry Groups Join Forces to Tell USDA to Reject Managed-Trade Proposal from Sugar Producers, American Jobs at Risk," December 4, 2009.

In: Sugar: Background, Policies and Issues
Editor: Cecilia M. Jordan

ISBN: 978-1-61761-116-2

Chapter 2

SUGAR POLICY AND THE 2008 FARM BILL*

Remy Jurenas

SUMMARY

Congress reauthorized the sugar price support program with some changes in the Food, Conservation, and Energy Act of 2008 (P.L. 110-246, the enacted 2008 farm bill). The sugar program is designed to guarantee the price received by sugar crop growers and processors and is intended to operate at "no cost" to the U.S. Treasury. To accomplish this, the U.S. Department of Agriculture (USDA) controls supply by limiting the amount of sugar that processors can sell domestically under "marketing allotments" and restricts imports. At the same time, USDA seeks to ensure that supplies of sugar are adequate to meet domestic demand. "No cost" is achieved if USDA applies these tools in a way that maintains market prices above minimum price support levels.

Since January 1, 2008, sugar imports from Mexico no longer face quotas or duties under the North American Free Trade Agreement. Other imports are allowed entry under quotas found in other free trade agreements (FTAs). To address the potential for a U.S. sugar surplus caused by additional imports under these trade agreements, the enacted farm bill mandates a sugar-forethanol program. USDA is now required to purchase as much U.S.-produced sugar as necessary to maintain market prices above support levels, to be sold to bioenergy producers for processing into ethanol. Funding is open-ended for this program. Other provisions increase the minimum guaranteed prices for raw sugar and refined beet sugar by 4% to 5%, mandate an 85% market share for the U.S. sugar production sector, and remove certain discretionary authority that USDA exercises to administer import quotas.

The enacted sugar provisions reflect the proposal presented to the House and Senate Agriculture Committees by producers of sugar beets and sugarcane and the processors of these crops. They favored continuing the structure of the current sugar price support program but sought changes to enhance their position in the U.S. marketplace. Their sugar-for-ethanol

* This is an edited, reformatted and augmented edition of a United States Congressional Research Service publication, Report RL34103, dated January 29, 2009.

provisions ensure that the prospect of imports adding to U.S. sugar supplies under any future trade agreements will not undermine the program's price guarantee and the sugar industry's market share. Food and beverage manufacturers that use sugar opposed the proposed program's provisions, arguing that costs to consumers will increase and that new requirements will restrict the flow of sugar for food use in the domestic market. The Bush Administration opposed these provisions, with the President identifying them as one reason why he vetoed the farm bill.

USDA has continued to estimate a tight domestic sugar supply in FY2009 largely due to reduced beet production. Its import quota decisions made to date and its estimate of sugar expected to enter from Mexico and other FTA partners do not point to a sugar surplus. As a result, USDA announced in September 2008 that the sugar-for-ethanol program will not be implemented this year. Attention now turns to how USDA will implement newly enacted rules dealing with the timing of additional raw cane sugar versus refined sugar imports, because of the implications for market prices.

Recent Developments

Current spot prices for refined beet sugar, as reported by a food industry trade journal,[1] are considerably above the loan forfeiture level (i.e., high enough so that there is likely to be no incentive for a beet processor this coming summer to hand over sugar pledged against a price support loan to the U.S. Department of Agriculture (USDA)). However, the raw cane sugar futures price for the July-September 2009 period, when such loans come due, continues to skirt around the loan forfeiture level. USDA by law must monitor this development to be prepared to possibly take steps to meet the statutory objective that the sugar program operate at no cost to the U.S. Treasury (for further explanation, see "Overview of Sugar Program" below).

On September 30, 2008, USDA detailed key provisions for the FY2009 sugar program to reflect the changes made by the 2008 farm bill. USDA announced the distribution of the beet and cane shares of the FY2009 overall allotment quantity (OAQ)—the maximum amount that sugar processors can sell domestically—among eligible companies. USDA also announced that based on current supply and demand conditions, there is no need to implement the sugar-to-ethanol program in FY2009.

Earlier, on September 9, 2008, USDA set the national OAQ for FY2009 and announced the FY2009 raw sugar and refined sugar import quotas. USDA's accompanying statement acknowledged that it expects the domestic market will require additional supplies of sugar during FY2009, and indicated that appropriate adjustments will be made to these two key parameters to ensure the availability of adequate supplies of sugar. See "Implementation" under "Controlling Sugar Supply to Protect Sugar Prices", and "Marketing Allotments" below, for details on both decisions.

Overview of Sugar Program

The sugar program is designed to guarantee the minimum price received by growers of sugarcane and sugar beets, and by the firms (raw sugar mills and beet refiners) that process these crops into sugar. To accomplish this, the USDA limits the amount of sugar that processors can sell domestically under "marketing allotments" and restricts imports. USDA is required to operate the sugar program on a "no-cost" basis. This means USDA must regulate the U.S. sugar supply using allotments, import quotas, and related authorities so that domestic market prices do not fall below minimum price levels. These are based on the loan rates for raw cane sugar and refined beet sugar set out in law, which USDA uses to derive effective support levels (see "Level of Sugar Price Support" below, for explanation). If the market price is below the support level when a sugar price support loan comes due, its "non-recourse" feature means a processor can exercise the legal right to forfeit, or hand over, sugar offered to USDA as collateral for the loan in fulfillment of its repayment obligation. Should this occur, USDA would record a budgetary expense, or outlay, for such a transaction.

This chapter focuses on the issues raised by the sugar program provisions in the House and Senate farm bills. Also, see the Appendix for a side-by-side comparison of the sugar provisions in the enacted 2008 farm bill with previous law and the House and Senate farm bill provisions.

Issues in 2008 Farm Bill Debate

Consideration of future U.S. sugar policy revolved primarily around four issues. These were where to set the level of minimum price guarantees to be made available to processors, how to use two tools to manage U.S. sugar supply, authorizing any sugar surplus to be used as a feedstock for ethanol, and accounting for projected program costs. Though industrial users of sugar in food and beverage products initially explored converting the sugar program to operate similar to the programs in place for the major grains, oilseeds and cotton, this policy option did not receive further attention.

Level of Sugar Price Support

USDA is required to extend price support loans to sugar processors that meet certain conditions on passing program benefits to the farmers that supply them with sugar beets or sugarcane. These loans are made at statutorily set loan rates,[2] and account for most of the effective support level made available to producers and processors. USDA is required to use its other tools to protect this price guarantee.[3] Loan rates for raw cane sugar have not changed since 1985; for refined beet sugar, since 1992. These minimum prices have guaranteed producers of sugar crops and the processors that convert these crops into sugar, a U.S. price that since the early 1980s has ranged from two to four times the price of sugar traded in the world marketplace.

The enacted 2008 farm bill will increase sugar loan rates by 4% to 5% by FY2012. Conferees split the difference between the House- and Senate-proposed rate increases and adopted the Senate approach that proposed to increase rates in stages each year. The loan rate for raw cane sugar would rise in quarter-cent increments from the current 18.0¢ per pound to 18.75¢/lb., beginning with the 2009 sugarcane crop. The refined beet sugar loan rate, beginning with the 2009 sugar beet crop, would similarly increase in stages, from the current 22.9¢ per pound to 24.1 ¢/lb in FY2012.[4] The Appendix provides loan rates for each of the fiscal years covered by 2008 farm bill authority.

Growers and processors had initially sought a one cent increase in the raw cane sugar loan rate (with a corresponding increase in the refined beet sugar rate), and had acknowledged their satisfaction with receiving half of their request in the House-passed farm bill. They argued that the increase in the loan rate is needed to cover increased production costs, particularly energy inputs. Sugar users countered that the House-proposed higher loan rates would increase costs to taxpayers by an additional $100 million annually. They also noted that while the bill's ethanol provisions (see "Sugar for Ethanol" below) "are supposedly designed to deal with surpluses," the loan rate increase "can only encourage *higher* surplus production."[5] The Bush Administration, in its statement of administration policy on the House and Senate farm bills, opposed the increase in the loan rates for sugar.

Implementation

On September 30, 2008, USDA announced loan rates for 2008-crop sugar as required by the 2008 farm bill. The national average loan rate is 18.0¢/lb. for raw cane sugar, and 22.9¢/lb. for refined beet sugar, the same as for the previous year's crop. In turn, these national loan rates are adjusted to reflect transportation cost differentials. Reflecting this, the raw cane sugar loan rate will range from 16.37¢/lb. in Hawaii (if sugar pledged for a loan is stored on the islands) to 18.22¢/lb. in Louisiana. The refined beet sugar loan rate will range from 21.95¢/lb. in Idaho, Oregon, and Washington to 24.34¢/lb. in Michigan and Ohio.

Controlling Sugar Supply to Protect Sugar Prices

The sugar program uses two tools—import quotas and marketing allotments—to ensure that producers and processors receive price support benefits. By regulating the amount of foreign sugar allowed to enter and the quantity of sugar that processors can sell, USDA can for the most part keep market prices above effective support levels, meet the no-cost objective, and ensure that domestic sugar demand is met. If these tools are implemented as intended, the likelihood that USDA acquires sugar due to loan forfeitures is remote.

Import Quotas

The United States must import sugar to cover demand that the U.S. sugar production sector cannot supply. However, USDA restricts the quantity of foreign sugar allowed to enter for refining and/or sale to manufacturers for domestic food and beverage use. Quotas are used to ensure that the quantity that enters does not depress the domestic market price to below support levels. Quota amounts are laid out in U.S. market access commitments made under World Trade Organization (WTO) rules and under bilateral free trade agreements (FTAs).

The sugar program authorized by the 2002 farm bill accommodated, or made room for, imports of up to 1.532 million short tons raw value (STRV) each year.[6] This import level is one of the four factors that USDA used to establish the national sugar allotment (called the "overall allotment quantity"), and reflected U.S. trade commitments under two trade agreements in effect when the 2002 program was authorized (**Table 1**).

Table 1. Annual U.S. Sugar Import Commitments When the 2002 Farm Bill Was Enacted.

		short tons
World Trade Organization Quota (minimum)[a]		1,256,000
North American Free Trade Agreement (NAFTA)—Mexico Quota (maximum)[b]		276,000
	Total	**1,532,000**

a. Covers both raw sugar and refined sugar.
b. Applied only through the end of calendar year 2007.

Since January 1, 2008, however, U.S. sugar imports from Mexico are no longer restricted. Under NAFTA, Mexico no longer faces any tariff or quantitative limit on the amount of sugar exported to the U.S. market. With this opening, though, imports could fluctuate from year to year for various reasons. First, the amount of Mexican sugar exported to the U.S. market will depend largely upon the extent that U.S. exports of historically cheaper high-fructose corn syrup (HFCS) displace Mexican consumption of Mexican-produced sugar. Surplus Mexican sugar, in turn, would then likely move north to the United States.[7] Second, Mexico's sugar output, though trending upward, does vary from year to year, depending upon weather and growing conditions. Mexican government policy also is to hold three months worth of sugar stocks in reserve and to allow sugar imports when needed to meet demand and lower prices.[8] Third, Mexican sugar prices in recent years have for the most part been higher than U.S. sugar prices. To the extent that this occurs in the future, the incentive for a Mexican sugar mill to export sugar north in search of a better price is reduced. Fourth, U.S. buyers' concerns about the quality of Mexican sugar may limit the amount that actually flows north in the next few years.

Also, the United States has committed under other existing and pending bilateral FTAs to allow for additional sugar imports.[9] Such imports in 2013, the fifth year of the sugar program authorized by the 2008 farm bill, could total from about 420,000 tons to 1.215 million tons *above* existing WTO and FTA trade commitments and *above* the maximum amount of sugar allowed to enter from Mexico in 2007. The wide range reflects two varying assumptions made to estimate by how much HFCS use in Mexico might displace sugar consumption in Mexico and create a surplus available for export to the U.S. market.[10]

Legislation

The sugar program provisions in the enacted 2008 farm bill did not directly address the issue of additional sugar imports. Instead, a new sugar-for-ethanol program is authorized to handle the price-related impact of such imports (Section 9001 in the energy title; see "Sugar for Ethanol" and "Sugar Program Costs" below). However, other provisions prescribe how USDA must now administer import quotas. To cover shortfalls (because of hurricanes or other disastrous events) in what domestic sugar processors can sell under allotments, USDA is directed to ensure that most imports enter in the form of raw cane sugar rather than refined sugar. While most permitted imports have historically entered in raw form, USDA decisions to allow large quantities of refined sugar to enter after the late 2005 hurricanes significantly affected the competitive position of cane refineries in Louisiana and Florida that process raw sugar. Unlike 2001-2002, when the Congress considered the last farm bill, most cane refineries are now a key part of vertically integrated operations owned by raw sugar processors and/or sugarcane producers. The 2008 farm bill's policy change is intended to ensure that these cane refineries (which process raw sugar into refined sugar) can more fully use their operating capacity. Also, limiting the entry of refined sugar enhances the position of the domestic beet sector to increase their sales of refined sugar.

Conferees, though, did not adopt provisions found only in the House-passed bill that would have directed USDA to regulate when and how much raw cane sugar imports are allowed to be shipped to U.S. cane refineries. While USDA announced shipping patterns in FY2003-FY2005, the impact of the hurricanes led to a decision not to follow this long-standing practice in FY2006- FY2008. USDA justified removing these restrictions because of "changes occurring over time in the domestic marketing of cane sugar." The House-passed provisions could be viewed as intending to increase the transaction costs for countries that export larger amounts of sugar to the U.S. market and to give a slight competitive edge to domestic processors with respect to buyers. Food and beverage firms opposed "micro-managing" the timing of imports, noting that the application of such rules will limit the ability of cane refiners to efficiently use their processing capacity and could lead to serious shortfalls at times in the amount of sugar supplied to the market.[11] In commenting on the House bill, the Bush Administration expressed concern over requiring shipping patterns for quota sugar imports. Also, several countries eligible to ship sugar to the U.S. market expressed concern that the proposed regulation of the flow of imports would run counter to U.S. trade commitments. Because of the concern expressed that prescribing how sugar import shipping patterns should be administered would open up the United States to challenges by sugar exporting countries in the WTO, these provisions were dropped in conference.[12]

Implementation

In line with the changes made by the 2008 farm bill, USDA on September 9, 2008, announced that the FY2009 raw sugar tariff-rate quota (TRQ)[13] will be set at 1,231,497 million STRV—the U.S. minimum access commitment for raw sugar imports under WTO rules. Relatedly, USDA announced the TRQ for refined and specialty sugars at 104,251 STRV. This amount is 80,000 ST higher than the U.S. minimum refined sugar TRQ (24,251 STRV), increased in order to meet U.S. demand for organic sugar not available from the domestic producing sector. Both announcements reflect the enacted 2008 farm bill's requirement that USDA set both the raw sugar and refined sugar TRQs at the minimum levels required by U.S. WTO trade commitments by October 1, 2008. USDA's accompanying statement acknowledged that it expects the domestic market will require additional supplies of sugar (i.e., imports) during FY2009, and indicated that appropriate adjustments will be made to these TRQ levels and the national marketing allotment level (see below) to ensure the availability of adequate supplies of sugar.[14]

The enacted 2008 farm bill limits USDA's ability to allow additional sugar imports under the TRQs established to meet U.S. WTO trade commitments. However, USDA is given some discretionary authority to exercise on this matter. Any decision to increase the raw sugar TRQ and/or the refined sugar TRQ before April 1, 2009, now requires USDA to first declare that an emergency sugar shortage exists because of "war, flood, hurricane, or other natural disaster" or another similar event as determined by the Secretary of Agriculture. USDA could interpret the law's discretionary language to determine that an emergency sugar shortage exists and allow for additional imports to the extent it determines that market prices will not fall below support levels and result in loan forfeitures. The sugar production sector likely will argue that there is no need for additional imports, pointing out that there is no physical shortage of sugar. Industrial sugar users likely will petition USDA to allow for additional refined sugar imports, pointing out that projected low ending stocks are keeping refined sugar prices considerably above the historical average. To date, USDA has not yet issued regulations detailing how such a determination would be made. In the interim, sugar crop producers/processors and major sugar users are expected to weigh in on what USDA under the Obama Administration develops as rules to detail what constitutes an emergency sugar shortage.

Marketing Allotments

In the 2002 farm bill, the domestic production sector accepted mandatory limits on the amount of sugar that processors can sell—known as marketing allotments—in return for the assurance of price protection. It viewed allotments as a way to try to capture any growth in U.S. sugar demand, and assumed that the then-U.S. sugar import quota commitments would continue without change (see "Import Quotas" above). The statute, however, stipulated that if (1) USDA estimates imports will be above 1.532 million short tons, and (2) that such imports would lead USDA to reduce the amount of domestic sugar that U.S. processors can sell, then USDA must suspend marketing allotments. Suspending allotments because of additional imports raises the prospect of downward pressure on market prices if most U.S. sugar demand is already met. If the additional imports were to cause the price to fall below support levels, forfeitures would occur and USDA would be unable to meet the no-cost requirement. Including the allotment suspension provision in the 2002 farm bill was designed to ensure that USDA not lose control over managing U.S. sugar supplies for fear of the consequences that could be unleashed (i.e., demonstrate its inability to implement congressional policy).

Legislation

Implementation of the 2002 farm bill's marketing allotment authority resulted in the U.S. sugar production sector's share of domestic food consumption ranging from a low of 73% in FY2006 to a high of 89% in FY2004. Concerned that their market share would decline as sugar imports increase under various trade agreements (see "Import Quotas" above), sugar producers and processors decided to pursue a different approach in formulating their proposal for the 2008 farm bill. Adopted by farm bill conferees, an important new provision guarantees that the domestic production sector always benefits from a minimum 85% share of the U.S. sugar-for-food market. USDA is now required to announce an "overall allotment quantity" (OAQ)—the amount of sugar that all processors combined can sell—that represents at least 85% of estimated domestic sugar consumption. This is intended to address the sector's objective that imports not displace the ability of U.S. sugar processors to sell more of their output in each successive year, to the extent that U.S. demand for sugar grows.

Implementation

On September 9, 2008, USDA announced that the FY2009 OAQ will be 8.925 million STRV. This complies with the new statutory requirement that USDA establish the OAQ at not less than 85% of estimated U.S. human sugar consumption (projected at 10.5 million STRV for FY2009).

The FY2009 OAQ level is considerably higher than USDA's latest estimate of FY2009 sugar production. Though cane sugar output is projected to increase by 4% over FY2008, beet sugar production is expected to be almost 11% lower because of spring 2008 weather problems in North Dakota and Minnesota, a major beet-producing region, and reduced planted beet acreage. As of mid-January 2009, USDA projected 2008/2009 U.S. sugar production at 7.8 million ST, 1.1 million ST below the OAQ that USDA announced last September. With the OAQ split between the beet and cane producing sectors using the percentage shares laid out in law, each sector is expected to fully market all of the sugar that USDA projects will be produced during FY2009 (Table 2).

USDA is currently projecting historically low ending stocks at the end of the 2008/2009 marketing year (i.e., September 30, 2009). This normally would imply higher than average wholesale sugar prices in the late summer period, but that is not the case across the board this year. The raw cane sugar futures price has been below the loan forfeiture levels for Florida and Hawaii since late October 2008, and below the loan forfeiture levels for Louisiana and Texas from mid-November to early December 2008. By contrast, current spot U.S. refined sugar (cane and beet) prices are well above the historical average. Various explanations for this price divergence are offered. Some point out that continued imports of raw-sugar-equivalent product from Mexico keeps downward pressure on the U.S. raw cane sugar futures price. However, USDA projects that imports from Mexico this year will be somewhat lower than last year. Others note that the loss of the production capacity of the cane sugar refinery in Savannah, Georgia, caused by an explosion a year ago, has reduced demand for raw cane sugar, keeping the raw cane sugar price lower than might be the case otherwise. Higher-than-average refined sugar prices are also attributed to the loss of refined product that this refinery would normally supply.[15]

Table 2. Comparison of National Sugar Allotment to USDA-Projected Sugar Production, FY2009.

		Overall Allotment Quantity	Estimated Production[a]	Shortfall (Allotment Deficit)
	share	*short tons, raw value*		
National	100.00 %	8,925,000	7,800,000	-1,125,000
Beet Sugar	54.35 %	4,850,738	4,225,000	-625,738
Cane Sugar	45.65 %	4,074,262	3,575,000	-499,262

Source: USDA, "USDA Announces Fiscal Year 2009 Sugar Program," Release No. 0226.08, September 8, 2008; USDA, *World Agricultural Outlook Board, World Agricultural Supply and Demand Estimates*, January 12, 2009.

a. As of January 12, 2009.

In the interim, any call made by sugar users for USDA to increase the end-of-year U.S. sugar supply (to meet their desire for lower sugar prices) is limited by what USDA can do under the enacted 2008 farm bill to increase imports (see "Implementation" under "Import Quotas", above). If USDA were to determine before the midpoint of the marketing year (April 1) that a sugar shortage exists, it would first have to reassign existing allotment deficits to imports of raw cane sugar (i.e., increase the raw cane sugar TRQ to accommodate the deficit amount). If a sugar shortage still existed after taking such action, USDA would only then be able to increase the refined sugar TRQ if two conditions are met and this second increase in supply will not result in sugar loan forfeitures. Before taking this second step, USDA must take into account that both the sales of domestic sugar and the refining capacity of domestic raw cane sugar have "been maximized." On or after April 1, USDA can only allow for imports of raw cane sugar to address an allotment deficit.

Taking this year's current price outlook into account, any USDA decision to allow for additional imports of raw cane sugar would likely depress raw cane prices. As a result, USDA could see some sugarcane processors forfeit on their loans in late summer 2009—a development that would run counter to the enacted farm bill's intent that USDA operate the sugar program at no cost. For this reason, USDA may be reluctant to allow such an increase, even though refined sugar prices are expected to remain high by historical comparison.

Sugar for Ethanol

Background

Sugar producers and processors have had an ongoing interest in exploring the potential for using sugar crops and processed sugar as a feedstock to produce ethanol (a gasoline additive). In the 2002-2003 period, they encouraged USDA to explore selling forfeited sugar stocks to corn-based ethanol processors. A few ethanol producers experimented by adding sugar to speed up the ethanol fermentation process, but the results were disappointing.

In 2005, Congress approved the Dominican Republic-Central American Free Trade Agreement (DR-CAFTA) that gives six countries increased access for their sugar to the U.S. market. During the debate, producers and processors sought a deal with the Bush Administration on a sugar-forethanol package. Their objective was to have this option available to divert additional sugar imports under DR-CAFTA whenever domestic prices fall below support levels.[16] With Congress mandating in 2005 that the use of renewable fuels be doubled by 2012,[17] some advocated that sugar be considered as a feedstock along with other agricultural crops and waste. Separately, Hawaii mandated (effective April 2006) that 85% of the gasoline sold must contain 10% ethanol. This requirement assumes that over time, the sugarcane produced on the islands will be available to use as the prime feedstock for ethanol.

If the cost of feedstock is excluded, producing ethanol from sugar cane can be less costly than producing it from corn. This is because the starch in corn must first be broken down into sugar before it can be fermented. This extra step adds to the cost of processing corn into ethanol, when contrasted to using sugarcane or processed sugar. Further, sugar cane waste (bagasse) also can be burned to provide energy for an ethanol plant, reduce associated energy costs, and improve sugar ethanol's energy balance relative to corn ethanol.

Brazil's success at integrating sugar ethanol into its passenger vehicle fuel supply has stimulated interest in exploring prospects for sugar-based ethanol in the United States. However, wide differences in sugar production costs and market prices in the two countries cause the economics of sugar-based ethanol to differ significantly. In investigating the economics of ethanol from sugar, USDA concluded that producing sugar cane ethanol in the United States would be more than twice as costly as U.S. corn ethanol and nearly three times as costly as Brazilian sugar ethanol.[18] Feedstock costs accounted for most of this price differential.[19] The USDA study showed that while sugar ethanol may be a positive energy strategy in such countries as Brazil, it may not be economical in the United States.[20]

Legislation

The enacted 2008 farm bill incorporates a proposal presented to the Agriculture Committees by the U.S. sugar production sector. The "Feedstock Flexibility Program for Bioenergy Producers" requires USDA to administer a sugar-for-ethanol program using sugar intended for food use but deemed to be in surplus. USDA will sell both surplus sugar that it purchases if determined necessary to maintain prices above support levels, and the sugar acquired as a result of loan forfeitures, to bioenergy producers for processing into fuel grade ethanol and other biofuel. Competitive bids would be used by USDA to purchase sugar from processors, at a price not less than sugar program support levels, which it would then sell to ethanol firms. USDA would implement this program only in those years where purchases are required to operate the sugar program at no cost. USDA's CCC would provide open-ended funding.

Because it would cost much more to produce ethanol from U.S.-priced sugar than from corn, this new program would require a considerable subsidy to operate as intended. The prime market for such sugar likely would be existing corn-based ethanol facilities close to sugar beet producing areas (e.g., the Upper Midwest) and new plants constructed in sugarcane-producing states (Hawaii and Louisiana). Producers of ethanol from corn in the continental United States, though, would likely need to adjust their fermentation process and/or invest in new equipment to handle sugar. As a result, they may not be as interested in purchasing sugar as a feedstock unless the price is significantly discounted further (e.g., requiring even more of a subsidy) to reflect the additional costs of processing sugar instead of corn. However, the availability of this subsidy could facilitate the development of the ethanol sector in Hawaii and partially reduce the islands' dependence on importing gasoline for its vehicle transportation needs.

As designed, this program will rely on U.S.-produced (rather than foreign) sugar. The amount that USDA decides to purchase would approximate its estimate of the extent that imports under trade agreements reduce the U.S. sugar price below support levels. Producers supported this provision, viewing it as an insurance policy for receiving the benefits of a guaranteed minimum price for sugar marketed for food use. Sugar users opposed this program "to ostensibly manage surplus supplies." In their July 13, 2007, letter to Members of Congress, they argued that this authority "will likely be used to short domestic markets, further restricting the availability of sugar for food use in the U.S. market." They characterized this approach as "wasteful of taxpayer resources" because sugar is not price competitive with corn as a feedstock, and will require large subsidies to ethanol producers "to induce them to accept the sugar." The Bush Administration opposed this sugar-for-ethanol component, commenting that it would not allow USDA to dispose of surplus sugar to end uses other than ethanol production, even if "those uses would yield a much higher return for taxpayers."[21]

Outlook

The current U.S. sugar market outlook (i.e., demand considerably above current supply, implying that market prices will be above loan forfeiture levels) suggests that, at present, USDA will likely be able to meet the program's no-cost directive without having to activate the new sugar-forethanol program in FY2009. USDA confirmed this with its September 2008 determination that no sugar will be available for this program, taking into account its forecast that sugar loan forfeitures are unlikely in FY2009. The status of this program in subsequent years will depend on whether U.S. sugar production returns to more normal levels, and on how sugar users (particularly in the beverage sector) in both the United States and Mexico respond to higher HFCS prices caused by (until recently) high corn prices. If HFCS prices are higher than Mexican sugar prices, the likely result will be a smaller displacement of Mexican-produced sugar by HFCS imports from the United States, and thus a smaller surplus available to be exported without restriction to the U.S. market. This reportedly did occur during FY2008.[22]

Sugar Program Costs

For the six years covered by the 2002 farm bill (FY2003-FY2008), USDA succeeded in operating the sugar program at no cost, as directed by law. Budget forecasts in early 2007 had projected that the sugar program, if continued without change, would cost from almost $700 million (Congressional Budget Office (CBO)) to about $800 million (USDA) over the FY2008-FY2012 five-year period. Projected outlays reflected estimates of the budgetary impact of additional sugar imports from Mexico and from other countries with additional access to the U.S. market for their sugar under bilateral FTAs. Each cost projection assumed that the additional supplies depress the domestic sugar price below price support levels, and lead processors to forfeit on a portion of their loans.[23]

Outlook

The policy changes enacted in the 2008 farm bill are intended to head off these potential costs and ensure that USDA can operate the program at no cost. However, estimating future budgetary impacts is difficult, considering that market conditions can change quickly and dramatically, and can differ significantly from historical experience. In its latest baseline budget projection (January 2009), CBO assumes that USDA essentially succeeds in operating the sugar price support program at no cost. It estimates outlays of $2 million in each of FY2009-FY20 12, or a total of $6 million over the farm bill's five-year period. For the sugar-for-ethanol program, which is an integral part of the sugar program but scored separately as a non-commodity activity, CBO estimates outlays of $325 million over the farm bill period. The latter estimate assumes that USDA sells sugar to bioenergy producers at a much lower price than the sugar program's minimum guaranteed price which USDA is required to pay for the surplus sugar purchased or acquired from domestic sugar processors.

APPENDIX. COMPARISON OF 2008 FARM BILL SUGAR PROGRAM PROVISIONS WITH PREVIOUS LAW AND HOUSE AND SENATE BILLS

Prior Law/Policy	House-Passed Bill (H.R. 2419)	Senate-Passed Substitute Amendment (H.R. 2419)	2008 Farm Bill (P.L. 110-246)
No Net Cost Directive			
Requires USDA to the maximum extent practicable to operate the sugar non-recourse loan program at no net cost by avoiding sugar forfeitures to the CCC. [7 U.S.C. 7272 (g), 7 U.S.C. 1359bb (b), *7 U.S.C. 1359cc (b)(2)*]	Retains current no net-cost requirement. [*Secs. 1301 and 1303(b)]*	Same as the House bill. *[Secs. 1501 and 1504(b)]*	Continues no-cost requirement found in prior law. *[Secs. 1401, 1403*] Requires USDA to operate sugar-for-ethanol program (in Energy title) to ensure this no-cost directive is met. *[Sec. 9001]*
Price Support Levels, Loans and Payments			
Sets raw cane and refined beet sugar loan rates at 18.0¢/lb. and 22.9¢/lb., respectively, through FY2008. Expands loan eligibility to in-process sugars and syrups at 80% of the applicable cane or beet loan rates. Makes nonrecourse loans available to processors that meet specified conditions. Sets 9-month repayment term for such loans. *[7 U.S.C. 7272 (a, b, d, e, f)]*	Increases raw cane sugar and refined beet sugar loan rates to 18.5¢/lb. and 23.5¢/lb, respectively, for FY2009 through FY2013. *[Sec. 1301]*	Increases raw cane sugar loan rate to 19.0¢/lb. by FY2013, in 1/4¢ increments beginning in FY2010, as follows: ¢ / lb. FY2009—18.00 FY2010—18.25 FY2011—18.50 FY2012—18.75 FY2013—19.00	Increases raw cane sugar loan rate to 18.75¢/lb. by FY2012, in 1/4¢ increments beginning in FY2010, as follows: ¢ / lb. FY2009—18.00 FY2010—18.25 FY2011—18.50 FY2012—18.75 FY2013—18.75 Sets refined beet sugar loan at 22.9¢/lb. in FY2009.

(Continued)

Prior Law/Policy	House-Passed Bill (H.R. 2419)	Senate-Passed Substitute Amendment (H.R. 2419)	2008 Farm Bill (P.L. 110-246)
		Increases beet sugar loan rate, beginning in FY2010, to be set at 128.5% of the raw cane loan rate in effect each year (e.g., reaching 24.42¢/lb. in FY2013), or as follows: ¢ / lb. FY2009—22.90 FY2010—23.45 FY2011—23.77 FY2012—24.09 FY2013—24.42	Starting in FY2010, sets beet sugar rate equal to 128.5% of the raw cane loan rate in effect (e.g., rising to 24.09¢/lb. by FY2012, or as follows: ¢ / lb. FY2009—22.90 FY2010—23.45 FY2011—23.77 FY2012—24.09 FY2013—24.09
		[Sec. 1501]	Continues other provisions found in
Authorizes CCC to accept bids from sugar processors to purchase USDA- owned sugar in conjunction with reduced production of new sugar crops. [*7 U.S.C. 7272 (g)]*	Continues in-kind authority. Stipulates that planted beets or cane diverted from production can only be used as bioenergy feedstock. *[Sec. 1301]*	Similar to the House bill. [*Sec. 1501]*	Continues in-kind authority and adds House/Senate provision. [*Sec. 1401]*

(Continued)

Prior Law/Policy	House-Passed Bill (H.R. 2419)	Senate-Passed Substitute Amendment (H.R. 2419)	2008 Farm Bill (P.L. 110-246)
USDA now pays storage rates of 8¢ per 100 lbs. for raw cane sugar and 10¢ per 100 lbs. for refined beet sugar that has been forfeited under the nonrecourse loan program. *[15 U.S.C. 714b & 714c; 7 CFR Part 1423]*	No comparable provision.	Requires (only through crop year 2011) USDA minimum storage payment rates of 10¢/cwt. and 15¢/cwt. on forfeited raw cane and refined beet sugar. *[Sec. 1503]*	Adopts Senate provision. *[Sec. 1405]*
Authorizes CCC to provide financing to processors of domestic sugar to construct or upgrade storage and handling facilities. *[Sec. 1402]*	No comparable provision.	Retains authority, but stipulates that loans shall not require any prepayment penalty. *[Sec. 1502]*	Continues prior law and adds Senate provision. *[Sec. 1404]*
Marketing Allotments and Allocations			
To avert loan forfeitures, requires USDA to limit the amount of sugar processors can sell each year. This is done through a national "overall allotment quantity" (OAQ) that is split between beet and cane sectors (54.35% and 45.65%, respectively), and then allocated to individual processors. The OAQ must accommodate WTO and NAFTA import commitments (1.532 million short tons). If imports are greater, USDA's authority to implement allotments is suspended. *[7 U.S.C. 1359aa, 1359bb, 1359cc, and 1359dd]*	Continues purpose and structure of marketing allotments and allocations, but changes some key provisions. Changes formula to require USDA to set OAQ at not less than 85% of estimated human food and beverage sugar use. Eliminates allotment suspension provision. *[Sec. 1303(a)-(d)]*	Similar to the House bill. *[Sec. 1504(a)-(d)]*	Continues marketing allotment authority and adopts House/Senate provisions that: —require USDA to set OAQ at not less than 85% of estimated U.S. human consumption, and —eliminate allotment suspension trigger. *[Sec. 1403(a)-(d)]*

(Continued)

Prior Law/Policy	House-Passed Bill (H.R. 2419)	Senate-Passed Substitute Amendment (H.R. 2419)	2008 Farm Bill (P.L. 110-246)
Directs USDA to reassign unused raw cane and beet sugar marketing allocations first to other cane states and beet processors, respectively; second to cane processors within each state; third to sales of sugar in CCC's inventory; and fourth to imports. *[7 U.S.C. 1359ee]*	Requires that any reassignment of unused cane and beet allocations to imports in the fourth step must be met by imports "of raw cane sugar." *[Sec. 1303(e)]*	Similar to the House bill. *[Sec. 1504(e)]*	Adopts House/Senate change to prior law. *[Sec. 1403(e)]*
Trade-Related Provisions			
In accord with a 1994 trade commitment, USDA sets an annual global sugar import quota of not less than 1.256 million short tons. USTR allocates the quota among eligible countries, and also administers preferential sugar import quotas for free trade agreement partner countries. Effective January 1, 2008, Mexico can ship duty free an unlimited amount of sugar to the U.S. market.	Makes no changes to import quota commitments found in various trade agreements and laws.	Makes no changes to import quota commitments.	Makes no change to current U.S. trade commitments.
Requires USTR in 2002-07 to reallocate unused country quota allocations to other quota-holding countries with sugar to sell. *[7 U.S.C. 1359kk]*	Repeals requirement for reallocating sugar import quota shortfalls. *[Sec. 1303(i)]*	Similar to the House bill. *[Sec. 1504(i)]*	Adopts House/Senate repeal provision. *[Sec. 1403(i)]*
USDA has discretion to increase the size of global raw cane and refined sugar import quotas when domestic sugar supplies are inadequate to meet U.S. demand at reasonable prices. [*Chapter 17, additional note 5, of the U.S. Harmonized Tariff Schedule; 19 CFR Part 2001, Subpart A*]	Requires USDA to set quotas for raw cane and refined sugar at the minimum level necessary to comply with U.S. trade agreement obligations. In cases of emergency sugar shortages,	Similar to the House bill. *[Sec. 1504(j)]*	Adopts House/Senate provision on setting initial import quotas at minimum levels and laying out

(Continued)

Prior Law/Policy	House-Passed Bill (H.R. 2419)	Senate-Passed Substitute Amendment (H.R. 2419)	2008 Farm Bill (P.L. 110-246)
	before April 1 of each marketing year, requires USDA to increase supplies first by reassigning allotment deficits to imports of raw cane sugar (i.e., increase the raw sugar quota), and second the refined sugar quota, if certain conditions are met. On or after April 1, allows USDA only to increase the raw cane sugar quota, if specified conditions are met. *[Sec. 1303(I)]*		steps to be followed to increase imports in the event of a sugar shortage. *[Sec. 1403(j)]*
To protect domestic sugar prices, USDA regulated the flow of sugar imports from large quota holders (through 2005).	Requires USDA to establish "orderly shipping patterns" for major suppliers of sugar to the U.S. market. *[Sec. 1303(i)]*	No comparable provision.	Deletes House "shipping patterns" provision.
The U.S.-Mexican agreement on bilateral market access for sugar and high-fructose corn syrup (HFCS) created an industry and government task force to address problems that might arise after the elimination of tariffs on sweeteners on January 1, 2008. *[Exchange of Letters between USTR and Mexico's Secretariat of Economy, July 27, 2006]*	No comparable provision.	Expresses sense of Senate that U.S. & Mexican governments should coordinate their sugar policies to be consistent with U.S. international commitments, to avoid disruptions of each	Deletes Senate provision.

(Continued)

Prior Law/Policy	House-Passed Bill (H.R. 2419)	Senate-Passed Substitute Amendment (H.R. 2419)	2008 Farm Bill (P.L. 110-246)
		country's sweetener markets (sugar and HFCS). *[Sec. 1505]*	
The U.S. withdrew from the International Sugar Organization (ISO) in 1992 because of opposition to the allocation of country contributions to ISO's budget.	Requires the Secretary of Agriculture to work with the Secretary of State to restore U.S. membership in the ISO within one year. [Sec. 1302]	Similar to the House bill. *[Sec. 1504]*	Adopts House provision. *[Sec. 1402]*
Sugar-for-Ethanol Program (Feedstock Flexibility Program)			
No comparable provision.	Requires USDA (for FY2008-FY2012) to purchase sugar from those firms that sell sugar (equal to the quantity of imports that USDA estimates exceeds U.S. food demand), and to resell such sugar as a biomass feedstock to produce bioenergy, in a way to ensure that sugar price support program provisions (see above) operate at no cost and avoid loan forfeitures. Requires USDA to use CCC resources, including "such sums as are necessary," to implement this new authority. *[Sec. 9013]*	Similar to the House bill. [Sec. 1501]	Adopts House/Senate provisions, and extends program by one year (FY2013). Prescribes how CCC-inventory sugar is to be disposed for this Program and other purposes, and allows for the sale of CCC-inventory sugar in the case of emergency shortages of sugar for food use. *[Sec. 9001]*

End Notes

[1] *Milling & Baking News*, January 27, 2009, 40.

[2] For sugar, the loan rate is the price per pound at which the Commodity Credit Corporation (CCC)—USDA's financing arm—extends nonrecourse loans to processors. This short-term financing at below market interest rates (e.g., 0.625% for loans taken out in January 2009) enables processors to hold their commodities for later sale.

[3] The loan rates alone do not serve as the intended price guarantee, or floor price, for sugar. In practice, USDA sets marketing allotments and import quota levels in order to support raw cane sugar and refined beet sugar at slightly higher price levels. Each price level takes into account the loan rate, interest paid on a price support loan, transportation costs (for raw sugar), certain marketing costs (for beet sugar), and discounts. These are frequently referred to as "loan forfeiture levels" or the level of "effective" price support.

[4] The loan rate for refined beet sugar reflects the requirement that it be set each year equal to 128.5% of that year's raw cane sugar's loan rate, beginning in FY2010.

[5] Letter to Members of Congress, from food and beverage companies and trade associations, and public interest groups, July 13, 2007.

[6] One short ton equals 2,000 pounds.

[7] However, the 2007 to mid-2008 increase in U.S. HFCS prices due to the higher cost of corn—its main input—did reduce its competitiveness against Mexican-priced sugar. To the extent that the HFCS price falls below sugar prices in the future, the incentive increases for Mexican bottlers of soft drinks to shift to HFCS.

[8] U.S. sugar processors also are now free to export sugar to Mexico to take advantage of the occasional higher prices there.

[9] All of the sugar access provisions in the Dominican Republic-Central American FTA (DR-CAFTA) already are in effect. Congress has yet to consider the FTAs with Panama and Colombia, which would grant additional access for each country's sugar into the U.S. market.

[10] The assumptions are laid out in an analysis that appeared in USDA's Economic Research Service's *Sugar and Sweeteners Outlook*, January 30, 2007, pp. 21-25.

[11] Letter to Members of Congress, July 13, 2007.

[12] The World Trade Organization administers trade dispute settle procedures whereby a country can file a case against another alleging that the latter operates a program or policy that runs counter to WTO rules. In this context, the prospect arose that a sugar exporting country might allege that the proposed shipping patterns provision were discriminatory or trade distorting.

[13] The quota component of a TRQ provides for duty-free access of a specified quantity of a commodity. Imports above this quota are subject to a prohibitive tariff.

[14] "USDA Announces Fiscal Year 2009 Sugar Program," September 9, 2008, as accessed at http://www.usda.gov/wps/ portal/!ut/p/_s.7_0_A/7_0_1OB?contentidonly=true&contentid=2008/09/0226.xml.

[15] Imperial Sugar's Savannah operations supply almost 10% of the refined sugar consumed in the United States. The firm plans to resume refining raw cane sugar to produce bulk granulated sugar in early 2009, and to completely restore its packaging facilities by the fall of 2009.

[16] Though the Administration did not agree to such a package, the Secretary of Agriculture pledged to divert surplus sugar imports—through purchases—for ethanol and other non-food uses, to ensure that the sugar program operates as authorized only through FY2008. For additional information, see "Sugar in DR-CAFTA—Sugar Deal to Secure Votes" in CRS Report RL33541, *Background on Sugar Policy Issues*, by Remy Jurenas.

[17] For more information, see CRS Report R40168, *Alternative Fuels and Advanced Technology Vehicles: Issues in Congress*, by Brent D. Yacobucci.

[18] Office of Economics, The Economic Feasibility of Ethanol Production from Sugar in the United States, July 2006.

[19] In Brazil, the cost of producing raw cane sugar reportedly ranges from 6¢ to 9¢ per pound (i.e., 9¢ to 12¢/lb. when converted to refined basis). In the United States, raw cane sugar production costs range from 12¢ to 20¢/lb. U.S. production costs for refined beet sugar range from 17¢ to 33¢/lb. For additional perspective, see "Costs of Production and Sugar Processing" in USDA, Economic Research Service, *Sugar Backgrounder*, July 2007, pp. 17-21.

[20] This discussion is adapted from "Sugar Ethanol" in CRS Report RL33928, *Ethanol and Biofuels: Agriculture, Infrastructure, and Market Constraints Related to Expanded Production*, by Brent D. Yacobucci and Randy Schnepf.

[21] Office of Management and Budget, "Statement of Administration Policy" on the Senate bill (Food and Energy Security Act of 2007), November 6, 2007, p. 3.

[22] Inside U.S. Trade, "USDA Projects Lower Mexican Sugar Exports; Corn Syrup Price Link Seen," June 6, 2008, pp. 11-12.

[23] The forfeiture of a price support loan results in a budget outlay, because the credit that had been extended is not paid back by the processor (resulting in a loss to the U.S. government). To the extent USDA succeeds in selling forfeited sugar, proceeds flow back to USDA and reduce the loss.

In: Sugar: Background, Policies and Issues
Editor: Cecilia M. Jordan
ISBN: 978-1-61761-116-2

Chapter 3

INDIAN SUGAR SECTOR CYCLES DOWN, POISED TO REBOUND*

Maurice R. Landes

ABSTRACT

A cyclical decline in sugar production is shifting India, the world's second largest producer, from net exporter to net importer during 2009/10 (October/September) and contributing to the current runup in global sugar prices. The downturn in production is primarily due to a policy-induced cycle that has become increasingly pronounced. India is forecast to shift from net sugar exports of 5.8 million tons in 2007/08 to net imports of 2.8 million tons in 2008/09 and a record 6.0 million tons in 2009/10. Sugar production is poised to rebound in 2010/11, as higher government price supports and open-market prices are likely to stimulate plantings and improve incentives to deliver sugarcane to sugar mills. In the longer term, India has the capacity to boost sugarcane output, and the government and the sugar industry are considering policy measures to moderate the increasingly sharp cycles in sugar production and trade.

ACKNOWLEDGMENTS

The author thanks Robert Knapp and Dr. Santosh K. Singh of USDA's Foreign Agricultural Service and Amani Elobeid of Iowa State University for their very helpful review comments, and John Weber and Wynnice Pointer-Napper of USDA's Economic Research Service for editing and designing this chapter.

* This is an edited, reformatted and augmented edition of a United States Department of Agriculture publication, Report SSS-M-260-01, dated April 2010.

Introduction

A cyclical decline in sugar production is shifting India, the world's second largest producer, from net exporter to net importer during 2009/10 (October/September) and contributing to the current runup in global sugar prices. The downturn in Indian production is primarily due to a policy-induced cycle that has become increasingly pronounced. Large domestic surpluses during 2006/07 and 2007/08 led to lower domestic sugar prices, weakened incen-tives for sugarcane growers, and sharply reduced plantings for the 2008/09 and 2009/10 crops. Indian domestic sugar prices have now increased sharply since early 2009 because of reduced supplies—and despite large government releases of sugar and liquidation of government-owned buffer stocks. India is forecast to shift from net sugar exports of 5.8 million tons in 2007/08 to net imports of 2.8 million tons in 2008/09 and a record 6.0 million tons in 2009/10.

Indian sugarcane and sugar production are poised to rebound in 2010/11. A combination of higher government price supports and open-market prices will likely stimulate plantings and also improve incentives to deliver sugarcane to sugar mills rather than to alternate uses. In the longer term, India has the capacity to boost both area and yields of sugarcane, and the government and the sugar industry are engaged in an ongoing dialogue on policy measures to moderate the increasingly sharp cycles in sugar production and trade.

Recent Supply and Demand Developments

Indian sugarcane and sugar production are historically cyclical, and the cycle appears to have become increasingly pronounced in the 2000s (figure 1; app. table 1). After soaring to 30.8 million tons in 2006/07—39 percent above the previous record set in 2002/03—sugar production dropped to an estimated 16.1 million tons in 2008/09 and is forecast at 17.3 million tons in the current 2009/10 year. All of India's sugar is produced from sugarcane, and the swings in production have been driven primarily by similarly large swings in harvested area of sugarcane (figure 2). Estimated area of 4.25 million hectares[1] in 2009/10 is about 17 percent below the record amount in 2006/07. Although more than 90 percent of sugarcane area in India is irrigated,[2] yields are affected by the share of area that is relatively high-yielding first-year growth[3] and, to a lesser extent, by variation in rainfall. Relatively small shares of first-year growth in 2008/09 and 2009/10, as well as below-normal monsoon rainfall in 2009, likely contributed to reduced average sugarcane yields in 2008/09 and estimated yields for 2009/10.

In contrast to the volatility of production, Indian sugar consumption has continued to expand relatively steadily due to rising per capita incomes and government interventions to adjust stocks, facilitate trade, and assure adequate monthly availability (see figure 1). Despite the sharp drop in sugar production in 2008/9 and continued low output in 2009/10, sugar consump-tion has remained relatively stable and on trend due to monthly releases of "free-sale" sugar into the open market and allocations of subsidized "levy" sugar. However, despite net imports of 2.8 million tons in 2008/09 and an anticipated 6.0 million tons in 2009/10, Indian sugar stocks are forecast to fall to 3.5 million tons by October 2010, the lowest level since 1993/94. Domestic open-market prices of sugar and gur,[4] the primary substitute sweetener, remain under pressure. For the 2008/09 marketing year, wholesale sugar prices averaged 43 percent higher in real terms than prices a year earlier, while gur prices averaged 53 percent higher (Figure 3).

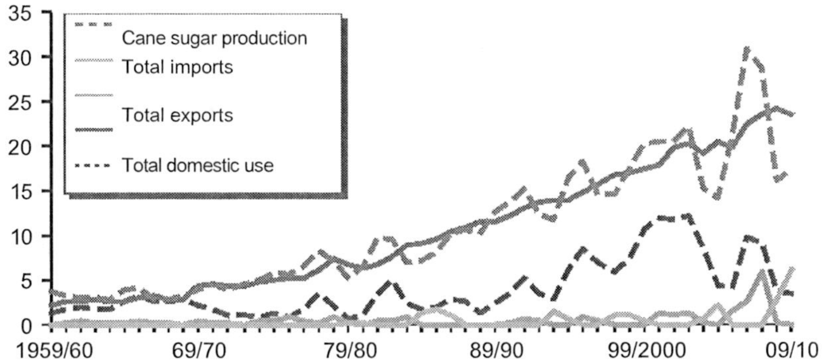

Source: USDA, Economic Research Service using data from USDA, Foreign Agricultural Service, PS&D Online.

Figure 1.Supply and use of centrifugal sugar in India (October/September marketing years).

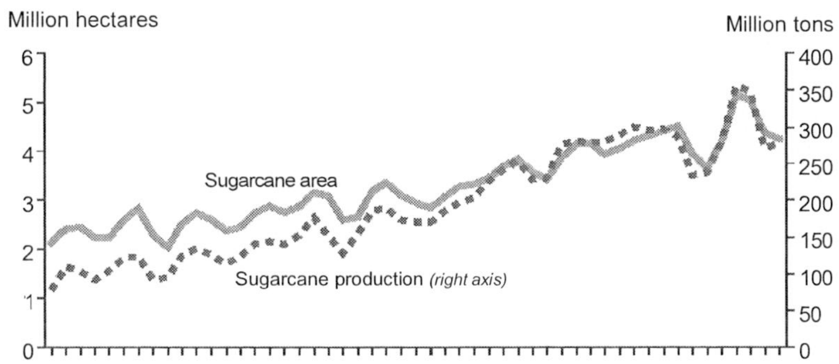

Figure 2. Sugarcane area and production in India.

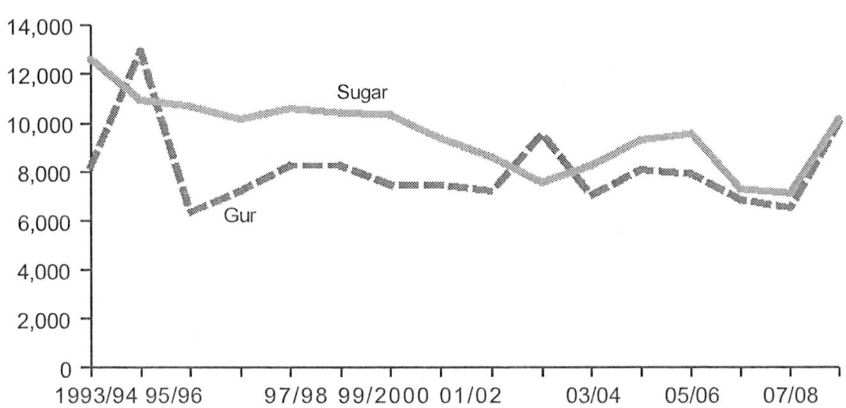

Source: USDA, Economic Research Service using data from Government of India, Ministry of Commerce and Industry, Office of the Economic Adviser.

Figure 3. Real wholesale prices of sugar and gur in India.

POLICIES DRIVE SUGARCANE PRODUCTION CYCLES

India's sugar area and production cycles are driven largely by policy inter-ventions, including sugarcane support price policies set by the central and state governments as well as sugar storage and trade policies set by the central government. Biological factors also play a role: sugarcane remains in the field for 3 years once it is planted, and area and production adjust down-ward slowly as price incentives fall, thus prolonging periods of oversupply, weak market prices, and financial distress for sugar mills. The key policy interventions are:

- Central and state government price support policies for sugarcane.
- Central government regulation of releases of levy and free-sale sugar and buffer stocks.
- Central government regulation of sugar trade.
- Other domestic marketing restrictions (e.g., private storage limits).

Central and state government price support policies for sugarcane. Under the system in place through the 2008/09 marketing year, the Commission on Agricultural Costs and Prices (CACP) under the Ministry of Agriculture recommended annual Statutory Minimum Prices (SMP) based primarily on estimated costs of production. State governments in some of the major producing states then set higher State Advised Prices (SAP) that mills in the state are required to pay sugarcane growers. The substantially higher SAPs set by some state governments account for regional variations in the productivity and profitability of sugarcane, as well as pressures from local sugarcane growers.[5] The central government financed the cost of supporting the SMP, but sugar mills were required to pay the difference between the SMP and the higher SAPs and incur any resulting financial losses. Begin-ning in 2009/10, the CACP will set Fair and Remunerative Prices (FRP) for sugarcane that, according to the new regulation, will also include "reasonable margins for growers of sugarcane on account of risk and profits."[6] With the new FRP system, the sugar mills gain some relief because more of the cost of supporting the current SAPs is shifted from the mills to the central govern-ment.[7] The future impact of the FRP system is unclear, however, and it is possible that it will encourage states to set even higher SAPs that are further removed from market conditions.

Data reveal a strong but lagged relationship between changes in the SAPs for sugarcane and changes in area harvested (figure 4). Drops in area harvested in 2003/04 and 2004/05 were preceded by declines in real SAPs, and higher SAPs in 2004/05 and 2005/06 corresponded with increased area in 2005/06 and 2006/07. Most recently, the drop in area in 2008/09 was preceded by sharply lower real SAPs in 2007/08. The fall in sugarcane area in 2008/09 was likely also infl uenced by the unusually large increases in Minimum Support Prices (MSPs) for wheat and rice, which compete with sugarcane for irrigated land, in 2006/07 and 2007/08 (Figure 5). Unlike the SMPs for sugarcane, the MSPs set by the central government for wheat and rice are good indicators of prices received by growers because a large share of the marketed surplus of wheat and rice is purchased at the MSPs.

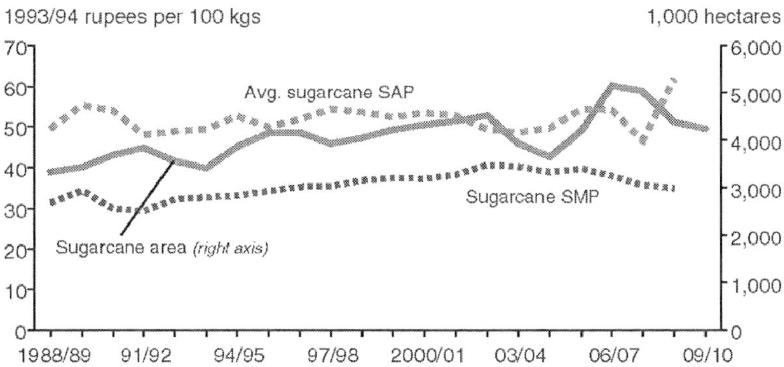

Sources: USDA, Economic Research Service using data from Government of India, Ministry of Agriculture, Directorate of Economics and Statistics; and USDA, Foreign Agricultural Service, Attache Reports.

Figure 4. Sugarcane area, real Statutory Minimum Price (SMP), and average State Advised Price (SAP).

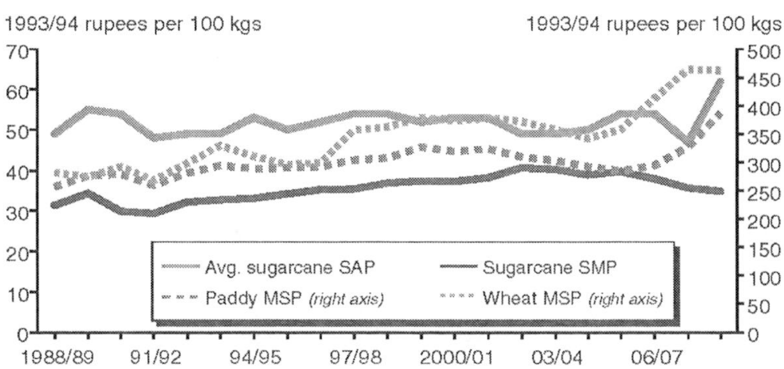

Sources: USDA, Economic Research Service using data from Government of India, Ministry of Agriculture and Cooperation, Directorate of Economics and Statistics; and USDA, Foreign Agricultural Service, Attache Reports.

Figure 5. Real State Advised Prices (SAP) and Statutory Minimum Prices for sugarcane and Minimum Support Prices (MSP) for paddy and wheat.

The SAP policy also affects the volatility of sugarcane area when it imposes financial losses on sugar mills that prevent them from honoring the SAP. During periods of surplus, SAPs can continue to rise without any specific link to sugar market conditions. These price hikes catch the mills between the fixed SAPs they are required to pay and weakening market prices they receive for their refined sugar. The resulting financial losses lead to deferral or default by mills on payments to growers that undermine incentives and contribute to volatility in sugarcane planting and production. Most recently, during 2006/07 and 2007/08, the large financial arrears to growers on the part of mills in major producing states reportedly led to both increased diversion of sugarcane from mills to production of khandsari[8] and gur, and reduced planting of new sugarcane area in 2008/09.[9] With higher open-market sugar prices in 2009, as well as central and state government measures to assist the mills, the finan-cial health of the mills has been largely restored and is not expected to affect processing or grower payments for the 2009/10 sugarcane crop.

Government regulated releases of levy, free-sale, and buffer stock sugar. The central government regulates all releases of refined centrifugal mill sugar into the market by sugar mills. Marketing of khandsari and gur, which are produced by farmers and small-scale enterprises, is unregulated. Mills are now required to sell 10 percent of their production at a fixed, below-market levy price to the Public Distribution System for sale to consumers determined to have incomes below the poverty line. The remaining 90 percent of mill production is sold at market prices, but the amounts that can be sold are determined by quarterly quotas set by the central government. Finally, buffer stocks accumulated by the government during periods of surplus are later released for open-market sale based on monthly quotas. A 5-million-ton buffer stock accumulated following record output during the 2006/07 marketing year was released during 2007/08 and 2008/09 to augment govern-ment supplies of levy and free-sale sugar.

Since 2000, the regulated releases of sugar by the central government have supported relatively steady growth in domestic supplies—and consumption—of centrifugal mill sugar (figure 6). The intent has been to reduce and eventually eliminate the requirement for mills to sell levy sugar at below-market prices. Although the mills have benefited from a reduction in the share of sugar they must sell at levy prices from 40 percent in 2000 to 10 percent by 2002, the government has not been able to deliver on the goal of eliminating the levy.

Government regulation of sugar trade. Trade in raw and refined sugar is tightly regulated through adjustments to import tariffs, export quotas, and, when needed, financial support to the industry to facilitate exports of surplus sugar. With the current shortfall in domestic supplies, the government, in February 2009, relaxed the re-export requirement for the Advanced License Scheme that allows mills to import raw sugar duty free subject to re-export of the equivalent volume of refined sugar. Later, the government announced unrestricted duty-free imports of raw sugar for April 2009-April 2010, temporarily eliminating the 60-percent basic ad valorem tariff, the Rs950/ton ($20/ton) countervailing duty, and the 3-percent education "cess" (tax) on the total tariff. Controls on imports of refined sugar were also eased by allowing duty-free imports of 1 million tons of refined sugar during August-November 2009. With these trade policy adjustments, 2008/09 sugar imports by India reached an estimated 2.8 million tons (consisting of about 2.5 million tons of raw sugar and 0.3 million tons of refined sugar), and 2009/10 imports are projected at about 6.0 million tons (consisting of about 5 million tons of raw sugar and 1 million tons of refined sugar).

Other restrictions on domestic marketing. The record rise in domestic sugar prices during 2008/09 led to additional regulatory measures aimed at providing short-term relief for consumers. In addition to the release of government-owned buffer stocks beginning in mid-2008, state govern-ments were empowered in March 2009 under provisions of the Essential Commodities Act to prevent "hoarding" of sugar by imposing stockholding limits and stock turnover requirements for privately held stocks with traders and food processors. Initially, the government imposed stock limits of 200 tons and a required turnover of 1 month, with some large government-recognized dealers and importers exempt from the restrictions. In August 2009, the government set maximum limits of 15 days of stocks for large bulk consumers.

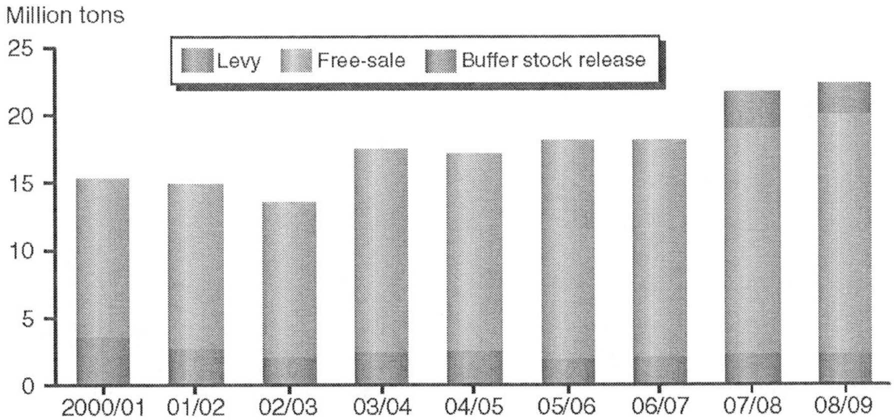

Source: USDA, Economic Research Service using data from Indian Sugar Mills Association.

Figure 6. Releases of levy, free-sale, and buffer stock sugarMillion tons.

In May 2009, the government suspended futures trading in sugar in an effort to control "speculative tendencies" in the market. This step was also taken in response to rising domestic prices of other commodities, including wheat and pulses, and refl ects official unease with the role and impacts of India's new futures exchanges in domestic commodity markets. The suspension of futures trading in sugar was initially intended to last through December 2009 but remained in effect as of March 2010.

Sugar Demand Trends

In contrast to the volatility of sugar production and prices, sugar consump-tion in India has grown relatively steadily, increasing about 3.7 percent annu-ally (2.0 percent per capita) since 1990/91 (figure 7). While policies have been associated with considerable volatility in sugar production, trade, stocks, and prices, government distribution programs have led to relatively stable growth in per capita sugar availability. The stable growth in consumption is also refl ective of both the robust growth in aggregate demand (outward shifting demand curves) and the price inelasticity, or unresponsiveness to changes in prices, of sugar demand. These factors stem from sugar's importance as an ingredient in tea, coffee, soft drinks, and traditional sweets that play a central role in Indian diets.

Refined centrifugal sugar from sugarcane is the dominant sweetener in India, with the two traditional sugarcane-based sweeteners, gur and khandsari, accounting for smaller shares of overall use. Khandsari has been declining in terms of production and consumption, and it now accounts for only about 3 percent of the market, but gur maintains a significant, albeit variable, 25-percent share of the market. Gur production and consumption are unregu-lated and tend to rise in years when higher gur prices or payment arrears by sugar mills create incentives for farmers to divert sugarcane to production of gur. During 2008/09 and 2009/10, gur's share of the market rose to about 35 percent when growers diverted sugarcane from the financially distressed mills, magnifying the drop in centrifugal sugar production caused by reduced sugarcane plantings (Figure 8).

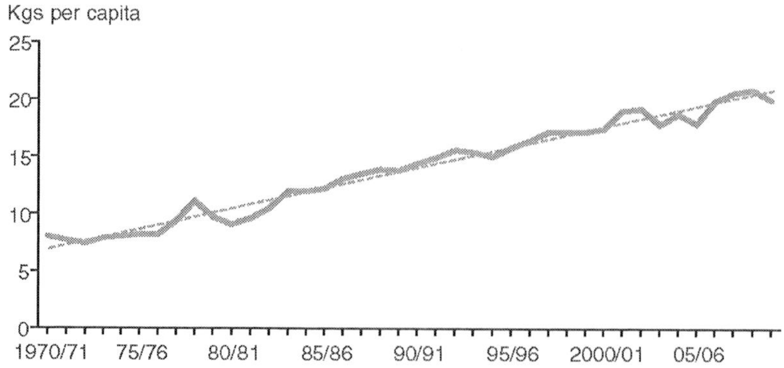

Source: USDA, Economic Research Service using data from USDA, Foreign Agricultural Service, PS&D Online.

Figure 7. Per capita sugar consumption in India.

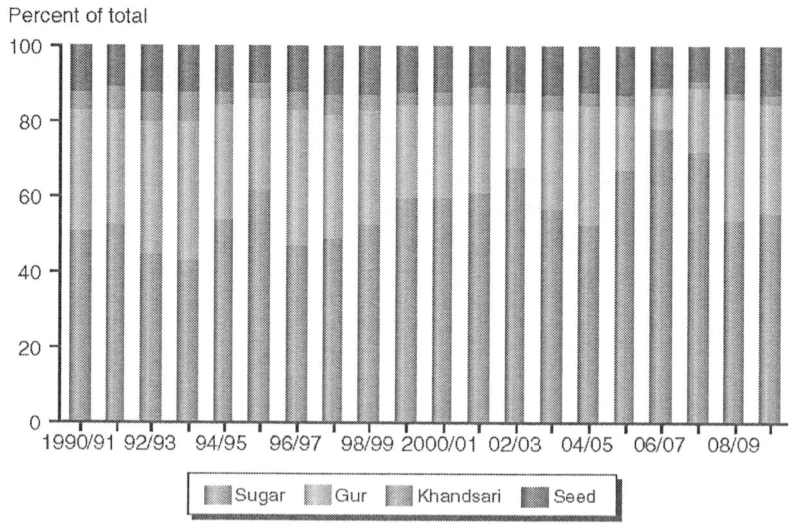

Source: USDA, Economic Research Service using data from Indian Sugar Mills Association.

Figure 8. Use of sugarcane in India.

Although India does have an ethanol-petrol blending program using ethanol produced from sugarcane, the program does not, so far, affect the supply of sugarcane for the production of sweeteners. Unlike Brazil, India produces all of its ethanol from molasses, which is a normal byproduct of India's centrifugal sugar milling process. As a result, there is no tradeoff between sugar and ethanol production. India's current goal is to require a 10-percent blend of ethanol with petrol in 20 states and 4 union territories, subject to its commercial viability. The drop in sugar production in 2007/08 and 2008/09 has led to reduced supplies and higher prices for molasses, and mills have been unable to deliver ethanol at the currently negotiated price of Rs21.5/liter (about $1.75/gallon). The ethanol blending program is expected to resume in 2010/11, assuming a recovery in output of sugar and molasses.

SUGAR TRADE

Like sugar production, sugar trade in India is cyclical, with exports of primarily refined sugar during periods of surplus and imports of mostly raw sugar during periods of deficit (figure 9). Consistent with the increased volatility of sugar production since the early 2000s, swings in sugar trade have also become more volatile. Since 2000, India's sugar trade has fl uctuated between average net exports of 1.3 million tons during 2000/01-2002/03, net imports of 1.2 million tons during 2003/04-2004/05, net exports of 3.3 million tons during 2005/06-2007/08, and forecast net imports of 4.3 million tons during 2008/09-2009/10.

The shifts in India's sugar trade are increasingly significant for world markets, contributing to periods of both undersupply and oversupply. India's record 2007/08 exports accounted for about 11 percent of global exports, and record imports in 2009/10 are forecast to account for 12 percent of world imports (figure 10). India's current shift to large net importer is further tight-ening a world sugar market that continues to adjust to European Union (EU) sugar policy reforms begun in 2006. The EU reforms, including reduced price supports, are due to be completed in 2009/10 and have led to sharp declines in sugar production and exports by member nations. During 2006/07 and 2007/08, the EU-25 averaged net sugar exports of 4.6 million tons, but by 2008/09, the EU-27 became a net importer of 2.1 million tons of sugar.[10]

The major sugar import markets affected by the instability of Indian supplies are mostly nearby countries in South and Southeast Asia and the Middle East, including Bangladesh, Sri Lanka, the United Arab Emirates, Pakistan, Malaysia, Indonesia, and Yemen (app. table 2). India is a small supplier of sugar to the U.S. market, with annual exports averaging 8,082 tons during 2000/01-2007/08. Brazil, the world's largest sugar exporter, is India's major supplier during years of deficit.

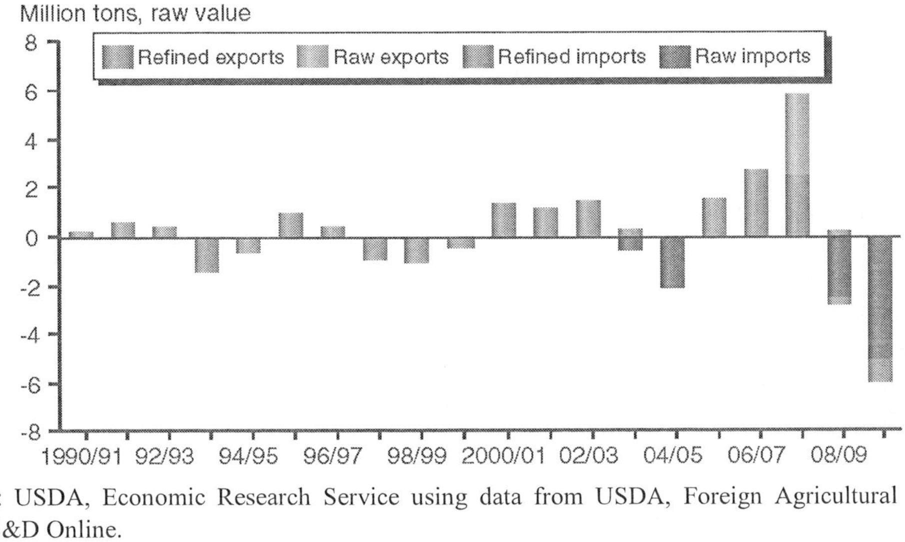

Source: USDA, Economic Research Service using data from USDA, Foreign Agricultural Service, PS&D Online.

Figure 9. Indian sugar trade.

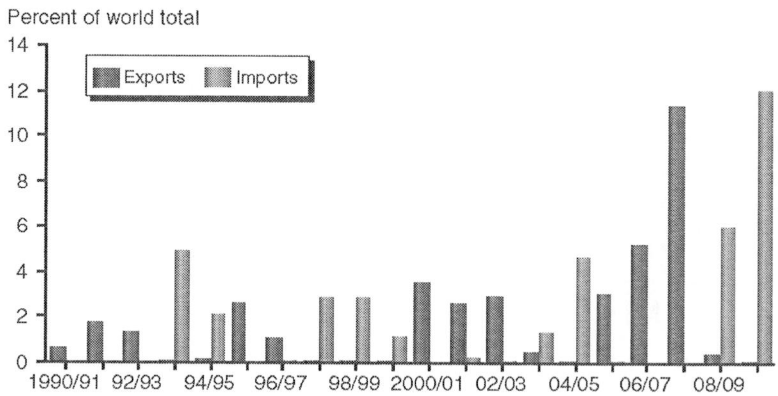

Source: USDA, Economic Research Service using data from USDA, Foreign Agricultural Service, PS&D Online.

Figure 10. India's share of world sugar trade.

Sugarcane Production Potential

India likely has significant potential to expand sugarcane production by increasing both planted area and yield. While India's area planted to sugar-cane, averaging about 4.5 million hectares per year of primarily irrigated land, is the second largest in the world (to Brazil), it accounts for a relatively small share of India's cropped area (about 142 million hectares) and net irrigated area (about 60 million hectares). Sugarcane, however, is a year-round crop that typically remains in the field for 3 years, and returns to sugarcane production must be competitive for irrigated land on which two, and in some areas three, crops are taken in one year. Figure 11 provides gross returns from sugarcane, wheat-paddy double crops common in irrigated areas of north India, and paddy-paddy double crops common in irrigated areas of south India over a 10-year period. A comparison shows generally higher returns to sugarcane, based on both average SAPs and the lower SMPs, although there has been a convergence in recent years due to the relatively large increases in wheat and paddy MSPs. However, cost of cultivation data indicate that labor costs for sugarcane are roughly double those for wheat and paddy, suggesting that labor availability and costs may be constraints to growth in sugarcane area.

India also appears to have the potential to improve sugarcane yields, and the average sugar content of harvested sugarcane, through a continued shift of planted area from northern states, where the climate is subtropical, to southern states, where the climate is tropical and conducive to higher sugar-cane yields and sugar recovery rates. Sugarcane yields in India average about 68 tons per hectare, about the same as China but below other major producers, such as Australia, Brazil, and Mexico (figure 12). But, an increasing share of India's sugarcane is being planted in southern states, where yields average about 83 tons/hectare, rather than in north India, where yields average about 58 tons/hectare (figure 13). Although the northern state of Uttar Pradesh—the historical center of the Indian sugar industry—still accounted for about 46 percent of sugarcane area and 39 percent of output during 2005/06-2007/08, the southern state of Maharashtra, where both average yields and the sugar content of sugarcane are higher, is now the largest producer of sugar. Most Indian sugarcane is grown under irrigation in all major producing states, providing favorable conditions for improving average yields (Figure 14).

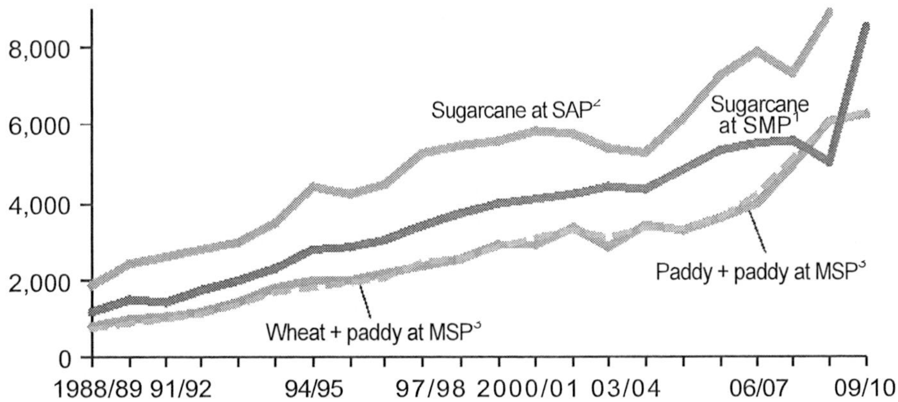

[1]Gross returns at Statutory Minimum Price.

[2]Gross returns at average State Advised Price.

[3]Gross returns at Minimum Support Price.

Sources: USDA, Economic Research Service using data from USDA, Foreign Agricultural Service, PS&D Online; and Government of India, Ministry of Agriculture, Directorate of Economics and Statistics.

Figure 11. Gross returns from sugarcane, wheat and paddy production in India.

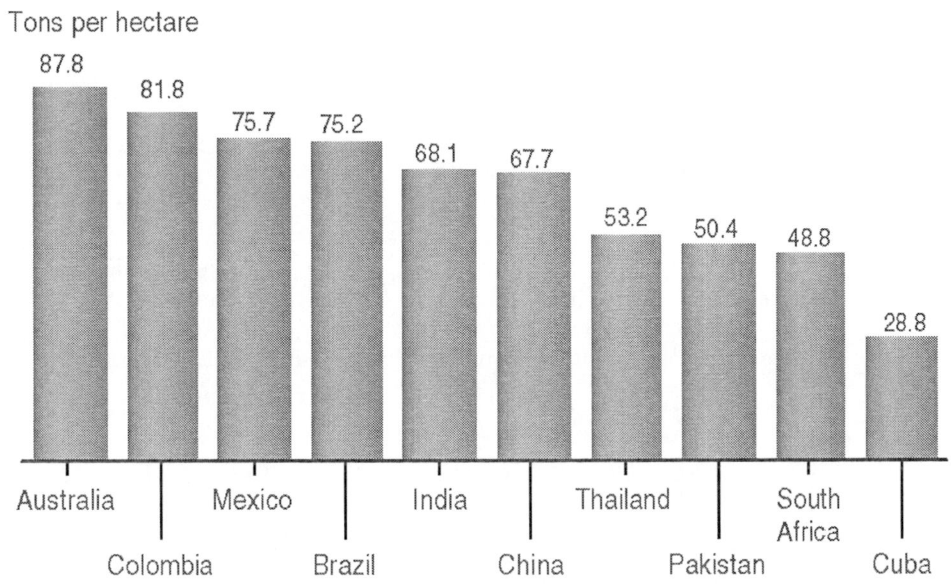

Source: USDA, Economic Research Service using data from Food and Agriculture Organization of the United Nations, FAOSTAT.

Figure 12. Top 10 world sugarcane producers ranked by 2005-07 average yield.

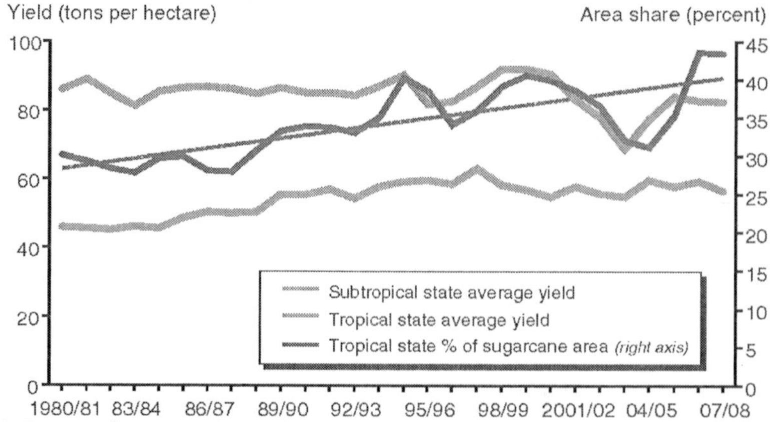

Source: USDA, Economic Research Service using data from Indian Sugar Mills Association.

Figure 13. Tropical and subtropical state sugarcane yield and tropical state share of area in India.

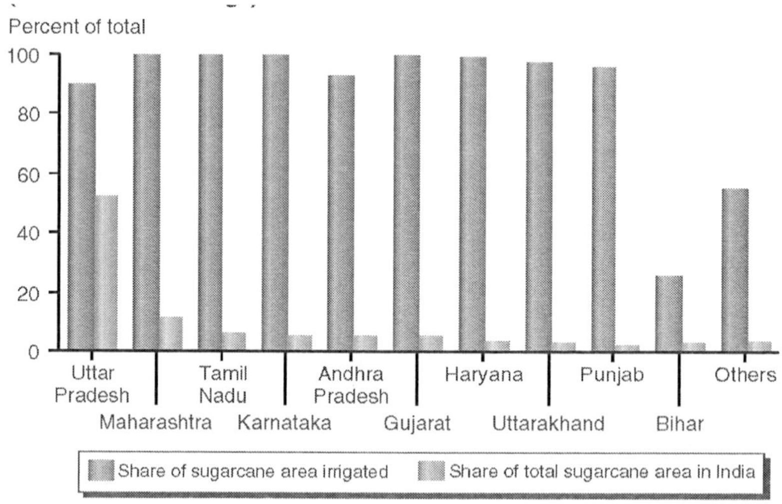

Source: USDA, Economic Research Service using data from Government of India, Ministry of Agriculture and Cooperation, Directorate of Economics and Statistics; and USDA.

Figure 14. Share of sugarcane area irrigated for major Indian states (2003/04-05/06 average).

Sugarcane and Sugar Output Poised for Rebound

With the 2009/10 Indian sugarcane harvest largely completed, sugarcane and sugar production will fail to recover from the poor 2008/09 outturn. Despite positive price signals, including a 32-percent real increase in the average SAP for sugarcane, the recovery in sugarcane area appears to have been impeded by financial distress in the sugar milling industry that led to large payment arrears to growers during 2008. A recovery in sugarcane yields was likely also slowed by a reduced share of relatively high-yielding first-year plantings, as well as shortages of irrigation water in some areas following the below-normal rainfall in 2009. With sugar stocks depleted by government releases of levy, free-sale, and buffer stock supplies in an effort to stabilize domestic prices, 2009/10 sugar imports are forecast at a record 6 million tons.

For 2010/11, the outlook is for a potentially strong recovery in sugarcane and sugar output that could sharply reduce or eliminate the need for addi-tional imports. With the financial condition of the sugar mills strengthened by high prices for free-sale sugar in 2009, growers may be more responsive to price support policies announced during 2009/10. Although the change in price policy triggered some initial uncertainty among growers, the new FRP for sugarcane for 2009/10—Rs130/100 kgs, 60 percent above the SMP announced for 2008/09—will transfer more of the burden of supporting higher SAPs from the sugar mills to the central government. So far, nominal increases in the SAPs for sugarcane have been announced in the major state of Uttar Pradesh (28 percent) and in Punjab and Haryana (21 percent). Smaller nominal hikes have been announced in the MSPs applicable in all states for paddy (11 percent) and wheat (2 percent). Sugarcane SAPs for southern states have not yet been announced for 2009/10.

On top of the large relative increases in sugarcane SAPs in 2008/09, the announced support prices suggest a strong recovery in sugarcane area in 2010/11. Most planting of new cane area will occur during March-May 2010 and will not necessarily be affected by the poor monsoon rainfall in 2009. With a return of normal weather patterns in 2010 and a surge in new plantings with relatively high yields, average yields are also likely to improve in 2010/11. A return to the average area (5.1 million hectares) and yield (69 tons/hectare) achieved during the most recent cyclical peak years of 2006/07-2007/08 would provide a sugar outturn of 29-30 million tons, likely sufficient to meet consumption and stock-building needs without significant additional imports. A return to longer term average area and yield would suggest an outlook for substantially lower production and higher imports in 2010/11. But, given the cyclical behavior of prices and output in the sector, the recent cyclical peaks may be a more suitable refer-ence for forecasting 2010/11 prospects.

The government and the sugar industry in India are engaged in an ongoing debate on potential policy reforms to stabilize the profitability of sugarcane production and processing and reduce the increasingly volatile cycles in the sector. There appears to be agreement that the key problem is a lack of alignment between administered prices for sugarcane on the one hand, and the administered and market prices for sugar and other byproducts on the other. Ideas that have been under discussion for a number of years include a shift to formula-based pricing that links sugarcane prices to prices for sugar and other byproducts, removal of the monthly release mechanism, establishment of strategic sugar stocks along with a price-band mechanism, and removal of nontariff restrictions on sugar trade. However, the two sides have not reached consensus on any specific reforms. The recent shift to the concept of supporting FRPs for sugarcane appears to provide states with the scope for setting SAPs for sugarcane that are further delinked from domestic and world market conditions, suggesting prospects for continued volatility.

Appendix Table 1. Production, supply, and use of sugarcane and centrifugal sugar in India (October/September marketing years).

	Sugarcane			Centrifugal sugar								
	Area	Yield	Production	Production	Imports			Exports			Domestic consumption	Ending stocks
					Raw	Refined	Total	Raw	Refined	Total		
	1,000 ha.	*Tons/ha.*	*1,000 tons*	*1,000 tons, raw value*								
1980/81	2,670	57.8	154,250	6,542	0	0	228	65	0	65	6,373	1,058
1981/82	3,190	58.4	186,360	9,727	0	0	0	408	0	408	6,823	3,554
1982/83	3,360	56.4	189,510	9,508	0	0	0	441	0	441	7,622	4,999
1983/84	3,110	56.0	174,080	7,042	0	81	81	765	0	765	8,900	2,457
1984/85	2,950	57.7	170,320	7,071	0	1,298	1,298	20	0	20	9,116	1,690
1985/86	2,850	59.9	170,650	8,149	0	1,732	1,732	39	0	39	9,493	2,039
1986/87	3,080	60.4	186,090	10,153	0	1,020	1,020	0	21	21	10,352	2,839
1987/88	3,280	60.0	196,740	10,591	0	76	76	0	19	19	10,886	2,601
1988/89	3,330	61.0	203,040	10,177	0	0	0	30	0	30	11,433	1,315
1989/90	3,440	65.6	225,570	12,575	0	53	53	0	32	32	11,535	2,376
1990/91	3,690	65.3	241,050	13,707	0	0	0	0	239	239	12,281	3,563
1991/92	3,840	66.1	254,000	15,249	0	0	0	0	590	590	12,977	5,245
1992/93	3,570	63.9	228,030	12,447	0	0	0	0	390	390	13,800	3,502
1993/94	3,420	67.2	229,660	11,704	0	1,500	1,500	0	30	30	13,900	2,776
1994/95	3,870	71.2	275,540	16,410	0	685	685	0	40	40	13,841	5,990
1995/96	4,150	67.7	281,100	18,225	0	0	0	0	940	940	14,820	8,455
1996/97	4,170	66.6	277,560	14,616	0	27	27	0	422	422	15,697	6,979
1997/98	3,930	71.1	279,540	14,592	0	1,000	1,000	0	21	21	16,700	5,850
1998/99	4,050	71.3	288,720	17,436	0	1,075	1,075	0	10	10	16,977	7,374
1999/00	4,220	70.9	299,320	20,219	0	438	438	0	25	25	17,296	10,710
2000/01	4,320	68.5	295,960	20,480	0	0	0	0	1,360	1,360	17,845	11,985
2001/02	4,410	67.4	297,210	20,475	100	0	100	0	1,130	1,130	19,760	11,670
2002/03	4,520	63.6	287,380	22,140	10	0	10	0	1,410	1,410	20,260	12,150
2003/04	3,930	59.5	233,860	15,150	550	0	550	0	250	250	19,115	8,485
2004/05	3,660	64.8	237,080	14,170	2,135	0	2,135	0	40	40	20,385	4,365

Appendix Table 1. (Continued)

	Sugarcane			Centrifugal sugar								
	Area	Yield	Production	Production	Imports			Exports			Domestic consumption	Ending stocks
					Raw	Refined	Total	Raw	Refined	Total		
2005/06	4,200	66.9	281,170	21,140	50	0	50	0	1,510	1,510	19,870	4,175
2006/07	5,150	69.0	355,520	30,780	0	0	0	0	2,680	2,680	22,425	9,850
2007/08	5,040	69.1	348,190	28,630	0	0	0	3,330	2,500	5,830	23,500	9,150
2008/09	4,380	61.9	271,250	16,130	2,500	300	2,800	0	190	190	24,200	3,690
2009/10	4,250	65.4	278,000	17,300	5,000	1,000	6,000	0	10	10	23,500	3,480

Sources: U.S. Department of Agriculture, Foreign Agricultural Service, PS&D Online; Government of India, Ministry of Agriculture and Cooperation, Directorate of Economics and Statistics.

Appendix Table 2. Indian sugar exports and imports by major partner (October/September).

Country	2000/01	2001/02	2002/03	2003/04	2004/05	2005/06	2006/07	2007/08	2008/09[1]
1,000 tons, raw value									
Exports									
Bangladesh	187.4	72.4	243.9	120.8	0.3	177.9	266.5	917.8	0.3
China	38.9	0.1	2.6	0.0	0.0	0.3	107.2	229.3	0.0
Djibouti	2.0	3.4	9.4	0.0	0.0	13.7	93.0	188.1	14.4
Egypt	0.5	0.0	49.6	0.1	0.0	0.0	33.0	206.8	0.0
Indonesia	196.3	166.1	114.3	24.0	2.0	38.3	200.6	175.8	0.3
Iran	5.0	12.8	5.4	0.0	0.0	0.0	51.0	221.5	0.6
Kenya	0.6	0.1	2.4	0.4	0.0	0.0	44.5	117.0	6.8
Malaysia	50.4	332.4	330.2	70.2	6.2	18.6	14.0	356.8	8.4
Pakistan	342.2	23.0	34.8	6.4	3.9	783.9	254.2	63.4	9.7
Saudi Arabia	2.5	0.4	0.7	0.6	0.9	0.7	25.4	245.9	19.4
Singapore	27.2	35.9	45.7	6.1	0.4	10.4	34.9	21.5	0.5
Somalia	9.5	15.1	101.4	16.2	1.6	27.7	152.4	292.1	7.2

Appendix Table 2. (Continued)

Country	2000/01	2001/02	2002/03	2003/04	2004/05	2005/06	2006/07	2007/08	2008/09[1]
Sri Lanka	152.5	278.6	354.7	100.1	15.4	149.0	309.7	549.7	71.3
UAE	32.9	63.8	72.5	12.5	1.8	20.3	94.1	1,289.2	23.5
United States	1.5	11.1	13.9	9.1	1.4	12.7	12.5	2.4	2.0
Yemen	22.6	8.0	146.8	0.2	0.1	26.3	346.0	269.9	1.4
Other	134.2	111.4	299.5	37.5	45.2	88.9	483.9	704.6	46.9
Total	1,206.4	1,134.7	1,827.7	404.1	79.2	1,368.7	2,523.1	5,851.7	212.7
Imports									
Australia	0.0	0.0	0.0	42.0	25.0	0.5	0.0	0.0	0.0
Brazil	0.0	0.0	4.7	206.9	1,075.8	0.1	0.0	0.0	985.7
Mozambique	0.0	0.0	0.0	0.0	21.0	0.0	0.0	0.0	0.0
Pakistan	33.9	0.3	0.0	0.0	0.0	3.6	0.0	0.0	0.0
South Africa	0.0	0.0	0.0	73.0	101.4	0.0	0.0	0.0	29.6
Other	0.6	0.0	0.6	2.2	11.8	3.3	0.5	0.2	73.4
Total	34.4	0.4	5.3	324.0	1,235.0	7.4	0.5	0.2	1,088.7

[1]October/August data.

Source: World Trade Atlas.

REFERENCES

[1] Food and Agriculture Organization of the United Nations. FAOSTAT. http://faostat.fao.org/default.aspx
[2] Government of India, Ministry of Agriculture and Cooperation, Directorate of Economics and Statistics. *Agricultural Statistics at a Glance*, various annual issues. http://agricoop.nic.in/Agristatistics.htm
[3] Government of India, Ministry of Agriculture and Cooperation, Directorate of Economics and Statistics. *Cost of Cultivation of Principal Crops in India*, 2007. http://dacnet.nic.in/eands/costofcultivation.pdf
[4] Government of India, Ministry of Agriculture and Cooperation, Commission for Agricultural Costs and Prices. *Reports of the Commission for Agricultural Costs and Prices for the Crops Sown During 2003-04 Season.*
[5] Government of India, Ministry of Commerce and Industry, Office of the Economic Adviser. Wholesale Price Index Data. http://eaindustry.nic.in/
[6] Government of India, Ministry of Consumer Affairs, Food, and Public Distribution. Sugar policy announcements. http://fcamin.nic.in/dfpd/ EventListing.asp?Section=Sugar&id_pk= 12 1&ParentID=0
[7] Government of India, Ministry of Consumer Affairs, Food, and Public Distribution, Department of Food and Public Distribution. Order S.O. 2665(E)/ EssCom./Sugarcane, New Delhi, October 22, 2009. http://fcamin.nic.in/ sugar/spf000 1 .pdf
[8] Indian Sugar Mills Association. 2006. *Indian Sugar Yearbook, 2005-06.*
[9] U.S. Department of Agriculture, Foreign Agricultural Service. Attache Reports.
[10] U.S. Department of Agriculture, Foreign Agricultural Service. Production, Supply, and Distribution Data Online.

End Notes

[1] One hectare equals 2.47 acres.

[2] The share of sugarcane area that was irrigated averaged 91 percent during 2003/04-2005/06, the most recent years for which data are available. During 1996/97-2005/06, the most recent 10 years for which data are available, the share of sugarcane area irrigated ranged from a low of 88 percent in 1995/96 to a high of 92.5 percent in 2005/06.

[3] Each planted hectare of sugarcane is typically maintained for three annual harvests: a relatively high-yielding first crop about 10 months after planting, followed by two lower yielding annual volunteer, or "ratoon," crops.

[4] Gur, or jaggery, is unrefined brown lump sugar made from sugarcane by farmers and unregulated, small-scale enterprises. It is a preferred sweetener for many rural consumers and is also used to produce illicit liquor.

[5] In southern states, average sugar-cane yields are higher, and average costs of production are lower, than in northern states.

[6] Government of India, Ministry of Consumer Affairs, Food, and Public Distribution, Department of Food and Public Distribution. Order S.O. 2665(E)/EssCom./Sugarcane, New Delhi, October 22, 2009. http://fcamin.nic.in/sugar/spf0001.pdf

[7] When the change in sugar support price policy was first announced in October 2009, the state governments, rather than sugar mills, were required to pay the difference between the new FRPs and the SAPs. With this shift in the financial burden, the major producing state of Uttar Pradesh announced a new SAP (Rs165/100kgs) that was *below* existing market prices, leading to political agitation by growers. In December 2009, the central govern-ment agreed to again make the sugar mills responsible for paying the differ-ence between the FRP and the SAPs. Uttar Pradesh then raised the SAP to Rs190/100kgs, with farmers still demanding that it be raised further to Rs205/100kgs.

[8] Khandsari is centrifugal sugar produced by small-scale, unregulated firms. Production is declining, and khandsari now accounts for about 3 percent of total sugarcane use.

[9] See "Sugar Demand Trends" on page 9.

[10] Bulgaria and Romania joined the EU in 2007, which accounts for the use of both "EU-25" and "EU-27."

In: Sugar: Background, Policies and Issues
Editor: Cecilia M. Jordan

ISBN: 978-1-61761-116-2

Chapter 4

THE EU SUGAR POLICY REGIME AND IMPLICATIONS OF REFORM*

Aziz Elbehri, Johannes Umstaetter and David Kelch

ABSTRACT

The European Union's sugar policy, in place since 1968, underwent its first major reform in 2005 in response to mounting and unsustainable imbalances in supply and demand. The reform, however, targeted only a few policy instruments (intervention price cut, voluntary production quota buyout, and restrictions on nonquota sugar exports), while leaving other key policies unchanged (interstate quota trading, sugar-substitute competition, and import barriers). Consequently, the extent of the reform's impact is limited, compared with more far-reaching alternatives, particularly when the oligopolistic nature of the industry and its noncompetitive pricing behavior are taken into account. A model- based analysis suggests that the reforms by themselves are unlikely to induce price adjustments sufficient to reduce overproduction unless quotas and/or high tariffs are reduced.

ACKNOWLEDGMENTS

The authors thank John Dunmore, USDA, ERS, for early support on the project and Mary Anne Normile, USDA, ERS, for carefully reviewing an early draft. Thanks for insightful feedback also go to Marinos Tsigas, International Trade Commission; Won Koo, North Dakota State University; William Janis, USDA, FAS; and Mark Gehlhar, John Dyck, Barry Krissoff, Linda Calvin, and Steven Zahniser, USDA, ERS. We are also grateful to John Weber for editorial expertise and Wynnice Pointer-Napper for layout and design.

* This is an edited, reformatted and augmented edition of a United States Department of Agriculture publication, Report 59, dated July 2008.

SUMMARY

The European Union (EU) is one of the leading sugar producers and traders in the world. This position was built over time through the application of protectionist policies that regulated production, prices, exports, and imports. Since its creation in 1968, the existing EU sugar policy—commonly referred to as the Common Market Organization (CMO) for sugar—had changed only marginally. In 2006, however, a new sugar regime took effect, largely influenced by three factors: tariff- and quota-free access to least developed countries (LDCs) beginning in 2009, which would likely result in increased sugar imports; enforcement of World Trade Organization (WTO) commitments; and the accession of 10 new member states to the EU in 2004, which exacerbated preexisting sugar supply-and-demand imbalances.

What is the Issue?

The EU sugar policy was reformed under multiple pressures, both internal and external. The reform package targeted a limited set of policy instruments, such as support prices and quotas, with the hope of improving the efficiency of the industry and making it more sustainable. However, the partial nature of the reform, which left several key policy interventions unchanged, raises concerns about the implications of the reform on the EU sugar industry and on international trade. Because the EU is the world's second largest producer and exporter of sugar and the third largest importer, the EU sugar reforms have important consequences for both global and U.S. sugar markets. This chapter examines the current EU sugar regime and uses a model-based approach to assess the potential market and trade implications of the implemented reforms.

What Did the Study Find?

Current sugar regime—The CMO for sugar is complex, encompassing a variety of policy instruments, including price support, production quotas (sugar and substitutes), export subsidies, and import barriers. The reforms targeted only a few of these instruments, principally cutting the intervention price, or the price guaranteed to EU producers. The 36-percent cut in the intervention price is designed to lower the market price and discourage sugar imports from LDCs. The reforms also included a voluntary buyout scheme for production quotas and a disallowance of exports of nonquota sugar, a step taken with the aim of reducing domestic production and bringing export subsidies within WTO limits.

The reforms did not address interstate quota trading (which could induce a significant shift in production from high- to low-cost regions), leaving in place, as before, national quota allocation. The reforms also retained production quotas on sugar substitutes (isoglucose, or high-fructose corn syrup), albeit at higher levels, preventing greater competition within the EU among different types of sweeteners.

Model-based results—According to a model-based analysis of the reform, the combined effect of cuts in prices and production quotas will lead to lower EU sugar production, lower prices for consumers, and higher consumption.

Moreover, when the prices of sugar substitutes go down, reducing the profit- ability of isoglucose production and, hence, its output, sugar consumption will rise. In all model scenarios, EU sugar exports decline as a result of a combination of lower production, lower export subsidies, and restrictions on exports of nonquota sugar. Consequently, EU sugar imports will have to rise to bring the market into balance.

When the model accounts for oligopolistic industry behavior, results show that EU sugar production will decline further than it would under the reform package's quota buyout scheme. Domestic market prices will not necessarily fall as much (or proportionally) as the cut in the intervention price, partly due to increased markups charged by sugar firms. Model results also show that EU sugar reform will lead to a significant exit of firms from the industry. This effect is consistent with the accelerated restructuring and consolidation of EU sugar processors that has been observed in the period following the reforms.

Market and trade implications—The analysis suggests that, given the market structure of the EU sugar industry, cutting the intervention price alone may not have the desired effect on production and market prices. The reform's impact on production and prices will likely depend more on how much sugar is actually removed from the market, which will be partly determined by the extent of the sugar industry consolidation currently under way. A more complete sugar sector reform that included lower import barriers (outside preferential agreements), interstate quota tradability, and greater market opportunities for sugar substitutes would have weakened oligopolistic behavior, thus having the potential to make the industry leaner and more competitive.

The reform's trade effects are more significant for preferential than for nonpreferential exporters to the EU. The near concurrent implementation of tariff-cutting measures in 2009 and the sugar reform in 2006 will dampen potential gains for LDCs with preferential access to the EU as they face greater competition from third-country (nonpreferential) exporters. Preferential exporters are likely to experience income loss (from preference erosion) despite possible increases in sugar exports to the EU. World sugar prices are expected to shift upward due to the EU sugar reform, but an even greater influence on prices may be rising demand for and production of sugar-based ethanol, mostly from Brazil. A tighter world sugar market and rising prices would benefit developing-country sugar exporters, partly offsetting export losses from the EU reform.

Finally, the EU sugar reform could have implications for the WTO. The reform not only reduces EU export subsidies to a level within current WTO limits, but it also might facilitate a phased removal of export subsidies by the EU as is called for in the Doha Development Round. In the area of domestic support, cuts in the intervention price and lower production would reduce the EU's aggregate measure of support. However, in the area of market access, the EU reform does not address the high import tariff barriers for sugar exporters outside preferential agreements. Reducing these import barriers could potentially have far greater impact on world sugar markets, including nonpreferential sugar trade.

How Was the Study Conducted?

This analysis of the EU sugar reform features qualitative and quantitative components. The qualitative assessment offers insights into the likely shifts in production, prices, and trade resulting from the reform, given key characteristics of the sugar industry. Due to the complexity of the EU sugar regime and the various interconnected issues to disentangle, the quantitative assessment of the reform required use of both partial and general equilibrium models. The partial equilibrium-based analysis incorporates details of the EU domestic sugar sector and policies, and the general equilibrium model directly addresses bilateral trade, distinguishing between preferential and nonpreferential trade flows and offering an account of both import and export policies. A unique contribution of this chapter is its explicit modeling of imperfect competition and the oligopolistic behavior of the EU sugar industry and how these attributes affect the expected responses to the policy reform.

INTRODUCTION

The European Union (EU) is one of the leading sugar producers and traders in the world. However, this position was built over time through the application of protectionist policies that regulated all aspects of the industry, ranging from production and prices to exports and imports. The existing EU sugar policy—commonly referred to as the Common Market Organization (CMO) for sugar—incorporates a set of market management tools, including an intervention price and production quotas. The CMO for sugar also includes export subsidies to dispose of excess supply, preferential trade arrangements for certain countries outside the EU, and high tariffs to insulate domestic producers from foreign competition.

Since its creation in 1968, the CMO for sugar has changed only marginally. In 1973, the United Kingdom (UK) acceded to the EU. As a result, the UK's preferential import commitments to its former colonies—the AfricaCaribbean-Pacific (ACP) countries—were incorporated within the CMO for sugar. In each of the attempts to reform the EU's Common Agricultural Policy (CAP)—including the 1992 MacSharry Reform, Agenda 2000, and the 2003 CAP reform—the sugar regime has virtually escaped reform.[1] But following the 2003 CAP reform, a combination of several internal and external factors created a greater sense of urgency to modify the sugar policy. First, the "Everything-But-Arms" (EBA) Initiative, in which the EU agreed to phase out tariffs by 2009 on imported raw sugar from 48 of the least developed countries (LDCs), opened the doors for additional preferential sugar imports. Second, a World Trade Organization (WTO) panel ruled that the EU sugar regime is in violation of the EU's WTO export commitments. This ruling requires the EU to stop out-of-quota sugar exports (which, according to the WTO, benefit from indirect subsidies through cross-subsidization) and include in the calculations of its export subsidies the amount for re-exported refined sugar made from raw imported sugar. To comply, the EU will be required to reduce its sugar exports and expenditures on export subsidies by a significant amount—from around 4 to 5 million tons per year to 1.3 million tons in volume and to 499 million euros in value. In practice, this reduction could make it no longer profitable for the EU to produce sugar outside the quota system, at least for food uses. Third, the enlargement of the EU from 15 to 25 countries (EU-15 to EU-25) in 2004 has exacerbated the EU's preexisting supply and demand imbalances of sugar.[2] While the 10 new member states from Central and Eastern Europe made a net contribution to EU sugar consumption, they also imported sugar from third countries (Australia, Brazil, Cuba, Mexico, and Guatemala). Thus, the enlarged EU-25 was required to compensate these third-country exporters under WTO rules with additional "current access" quotas,[3] estimated at around 490,000 tons (Huan-Niemi, 2003).[4] Together, these factors made a significant overhaul of the EU sugar regime almost inevitable.

A comparison of expected supply and demand projections over the short term reveals the severity of the market imbalance. Excess sugar supply, already chronic in the EU, is expected to worsen under the combined effect of export reductions resulting from EU compliance with the WTO ruling and implementation of the EBA Initiative. According to per year estimates from the Commission of the European Communities (2005), domestic sugar consumption (15.2 million tons) plus WTO-permissible exports (1.3 million tons) would fall far short of absorbing the supply from domestic production (19.8 million tons) plus agreement-bound preferential sugar imports (1.6 million tons), unless the EU modified its sugar regime. After accounting for high-end estimates of EU sugar imports from the EBA countries of around 3 million tons, excess supply could have reached 7-8 million tons had the EU not changed its sugar policies.

Beginning in 2005, the EU embarked on a reevaluation of its sugar policies. On July 14, 2004, the EU Commission published an initial proposal and modified it following the unfavorable WTO ruling on EU export subsidies for sugar. Member states reached a final agreement on a new sugar policy regime on November 24, 2005. The reforms were ratified by the EU Parliament on January 19, 2006, and later by the Council of Ministers on February 22, 2006. The new sugar policy entered into effect on July 1, 2006. This chapter examines the current EU sugar regime and uses a model-based approach to assess the potential market and trade implications of the implemented reforms.

OVERVIEW OF THE COMMON MARKET ORGANIZATION FOR SUGAR

The main objective of the CMO for sugar is to ensure sugar production in all member states where sugar beet production is feasible, regardless of cost competitiveness. National production quotas given to sugar processors based on historical production levels serve to accomplish this goal. The EU also operates an intervention price—a price guaranteed to producers—as the basis for intervention purchases. Moreover, it shields sugar from competition via several mechanisms. First, the EU imposes restrictive quotas on imports of sugar substitutes, such as isoglucose; high import tariffs protect against foreign competition. Export subsidies help to dispose of excess sugar supply and maintain high domestic prices. Also, sugar-using industries receive subsidies to compensate for the high price of sugar.

The EU-25 harvested 2.3 million hectares of sugar beets in 2005, with an average yield of 61.6 tons per hectare (Bureau et al., 2007). Productivity of sugar beet production varies signifiicantly across the EU (Figure. 1). The number of sugar beet processing plants in the EU- 15 shrank from 240 in 1990 to 135 in 2002/03 (table 1), while the number of sugarcane refineries fell to 6. With the EU's 2004 enlargement, about 100 sugar processing plants from the 10 acceding countries were added to the EU sugar sector. But by 2005, the total number of plants declined to 184, and sugar production totaled about 21 million tons. The EU-25's first and second largest sugar producers, Germany and France, account for about half of the EU-25's production, followed by Poland, Italy, and the UK.

Production Quotas

For 2005, the total quota amount for sugar production in the EU-25 is 17.4 million tons. The sugar quota allocated may be transferred between processors within member states but not between member states.

While the quota allocation can be revised yearly, in practice it is changed only in the case of the closure or merger of sugar beet processors. Cane refineries procure raw sugar through imports from third countries or overseas territories of the member states. Production quotas have changed little over the years, even though production began to exceed consumption in the EU around 1977. Production quotas were cut for the first time in 2000, and this cut was 3.3 percent.

A distinctive feature of the CMO for sugar, compared with CAP regulations for other agricultural products, is that the policy instruments target the sugar beet processors and sugarcane refineries, not the sugar beet growers. Processors procure sugar beets from growers on the basis of intratrade agreements defined by the CMO for sugar. These contractual agreements define standard sugar and beet qualities and set purchase terms for beets and rules for quota transfers between enterprises. Based on the quota allocated to sugar beet processors, the processors, in turn, assign "beet delivery rights" to individual growers as specified in the intratrade agreements.

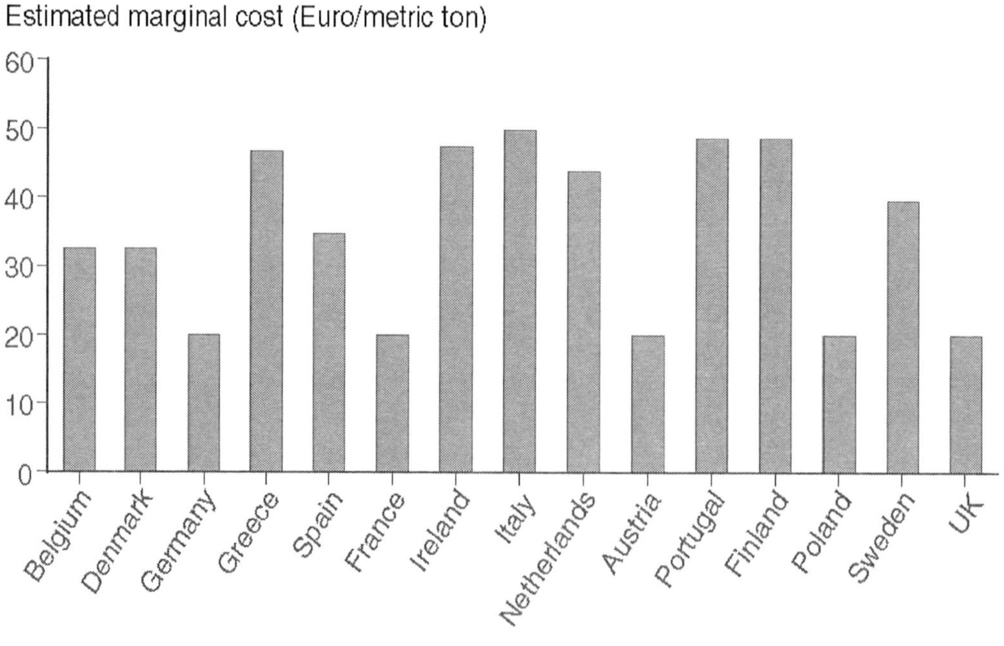

Source: USDA, ERS using data from Frandsen et al. (2003).

Figure 1. Estimated marginal cost for sugar across EU member states, 2003.

Table 1. EU sugar industry, 2002-03.

Country	Factories	Companies	White sugar yield	Sugar beet yield (t/ha)	Production + carryover	A sugar	B sugar	C sugar
	Number		Tons/ha		1,000 tons			
Austria	3	1	10.3	68.4	454.0	314.0	73.0	67.0
Belgium	8	6	10.4	66.0	888.0	675.0	145.0	68.0
Denmark	3	1	9.3	57.1	501.0	325.0	96.0	80.0
Finland	2	1	5.3	38.6	166.0	133.0	13.0	20.0
France	34	17	11.6	77.5	4,105.0	2,789.0	752.0	564.0
Germany	28	7	8.8	58.9	3,798.0	2,613.0	804.0	381.0
Greece	5	1	7.2	73.1	370.0	289.0	29.0	52.0
Ireland	2	1	6.3	52.3	223.0	181.0	18.0	24.0
Italy	20	9	5.7	51.8	1,563.0	1,311.0	162.0	90.0
The Netherlands	5	2	9.6	60.0	953.0	684.0	180.0	89.0
Portugal	4	4	9.0	62.5	56.0	56.0	0.0	0.0
Spain	13	3	10.4	76.8	1,086.0	957.0	40.0	90.0
Sweden	2	1	8.0	48.5	427.0	335.0	33.0	59.0
United Kingdom	6	2	9.7	64.7	1,260.0	1,035.0	104.0	121.0
EU-15 countries	135	56	8.7	61.8	15,850.0	11,697.0	2,449.0	1,705.0
Czech Republic	12	8	6.7	44.0	474.9	441.2	13.7	20.0
Hungary	6	5	5.4	40.0	401.7	400.5	1.2	0.0
Lithuania	4	2	4.9	35.0	112.0	103.0	0.0	9.0
Latvia	2	2	4.9	34.0	66.5	66.4	0.1	0.0
Poland	54	57	6.6	40.0	1,772.5	1,580.0	91.9	100.6
Slovakia	5	4	5.0	32.0	207.5	189.8	17.7	0.0
Slovenia	1	1	5.3	45.0	53.0	48.2	4.8	0.0
EU-25 total	219	135	8.3	58.2	18,938.1	14,526.1	2,578.4	1,834.6

Source: USDA, ERS using data from EU Commission and from Comité Européen des Fabricants de Sucre (CEFS).

Quotas comprise two quota types: "A-quota" (82 percent of total quota allocations), and supplemental "B-quota" (18 percent) to cover year-to-year production variability due to weather or unexpectedly low yields. Like the U.S. sugar program, the CMO for sugar is designed to be self-financed (i.e, it falls outside the EU budget), so a levy is imposed on sugar quota production. A-sugar is charged 2 percent of the intervention price (12.60 euros per ton of sugar); B-sugar is charged a variable levy determined by the total cost of export refunds but has a maximum level set at 37.5 percent (237 euros per ton) of the intervention price.

Sugar beet processors pay the levies, though they, in turn, pass on a portion of the cost to growers (Commission of the European Communities, 2005). If these levies prove insufficient to cover the costs of the export subsidy, an additional levy is imposed as a percentage of the amount owed by each processor under the previous two levies, with no maximum set to achieve the sum required.

Sugar produced outside the quota is called C-sugar (nonquota sugar). The price guarantee applies only to quota sugar (A and B). Hence, nonquota sugar is ineligible for any price support, nor can it be sold on the domestic market. C-sugar may be exported, but it cannot receive export subsidies. A carryover mechanism is available for C-sugar, which involves storing the sugar for a minimum period of 12 months (i.e., "carried over" to the next marketing year). Carried-over C-sugar is treated as A-sugar in the following marketing year, but the carryover amount of C-sugar cannot exceed 20 percent of the A-sugar quota (see table 1).

Sugar beet processors in member states that produce C-sugar average around 3 million tons of C-sugar per year, most of which is exported. Even though C-sugar receives no direct export subsidies, it is sold profitably at world market prices because the prices obtained for sales of A- and B-sugar (export subsidy inclusive) are high enough to cover all the fixed costs of the processing companies. This cross-subsidization (where C-sugar holders benefit indirectly because of the subsidies they receive on A- and B-sugar) was a key issue raised against the EU in the 2005 WTO ruling on export subsidies. According to the WTO Appellate Body, total subsidized sugar exports by the EU, including C-sugar, amount to almost 4 million tons per year, which far exceed the EU's WTO commitment limit of 1.273 million tons per year. With ACP sugar re-exports added to its total, the EU spends nearly 1.3 billion euros per year on export subsidies, more than twice the official ceiling (and WTO notified amount) of 499 million euros from 2000 to 2001 and later (Agra Europe, 2004).

The CMO does not cover sugar produced for certain industrial uses, and non-CMO sugar does not qualify for any CMO support. Consequently, there is no limit on non-CMO sugar production. This provision applies to sugar processed into alcohol, including fuel ethanol, rum, and spreadable syrups, and sugar used to produce yeasts. Currently, France is the EU's leading producer of sugar for nonfood purposes. In 2004, France reported harvesting about 40,000 hectares of sugar beets annually for nonfood purposes, of which about 10,000 hectares were used for ethanol extraction (Bruhns et al., 2004).

Sugar Substitutes

In 1977, the CMO for sugar was extended to production of isoglucose syrup to limit the product's competitive effect on the protected sugar industry. Isoglucose, also referred to as high-fructose syrup (HFS), is extracted from maize, wheat, and starch potato and is subject to an annual production quota of 500,000 tons for the EU-25. Unlike quotas for sugar, quotas for isoglucose are not eligible for purchase among member states; hence, they have no intervention prices. Since 1994, inulin syrup, another sweetener covered by the CMO for sugar, has been subject to a 300,000-ton per year quota (for the EU-15). Inulin syrup is treated the same as isoglucose in terms of quota restrictions but is more expensive to produce.

The cost of producing isoglucose in the EU has always been lower than the cost of producing sugar. A recent UK study shows that the production cost of isoglucose varies from 350 to 380 euros per ton, depending on the type. This cost range compares favorably with the breakeven price for sugar (below which it is no longer profitable to produce), which ranges from 300 to 350 euros per ton in some "low-cost" regions of France to 500 to 550 euros per ton in Italy (Commission of the European Communities, 2005). As early as 1995, the price of isoglucose was about 13-percent lower than the price of sugar in the EU, and the price is consistently below the intervention price of 631.90 euros per ton for white sugar. Given this price advantage, the principal factor that prevents a major substitution of isoglucose for sugar is the tight production quota on isoglucose. Nevertheless, in the EU, isoglucose prices follow sugar prices very closely and, therefore, would be affected by any reform to sugar prices or quotas. By contrast, inulin is a poor substitute for sugar, because inulin syrup is more expensive to produce (Netherlands Economic Institute, 2000).

Intervention Prices and Purchases

The EU maintains a minimum price for sugar via a two-tier strategy featuring a minimum price for sugar beets and an intervention price for white refined sugar. Intratrade agreements between sugar processors and beet growers regulate the relationship between the minimum sugar beet price (paid to growers) and processed sugar prices. Under the CMO for sugar, the basic beet price is the key support price. In 2005, the basic beet price was 4.767 euros per 100 kilograms. The basic beet price is a function of the intervention price for sugar, and its derivation takes into account the processing margin, the yield of sugar from beets, the revenue from molasses sales, and the cost of beet delivery to processors (table 2). Sugar processing returns (if sugar is sold into intervention) are split approximately 58 percent to sugar beet growers (i.e., 36.6 euros per 100 kilograms at the present intervention price) and 42 percent to processors (26.5 euros per 100 kilograms). The growers' share of 36.6 euros per 100 kilograms of sugar corresponds to the basic beet price of 4.767 euros per 100 kilograms on the basis of 7.7 tons of standard quality sugar beets (16 percent sugar content) being needed to produce 1 ton of sugar.

From the "basic price" of sugar beets is derived the minimum price *actually* paid to beet producers. Sugar processors are required to pay the minimum price, which is the basic beet price minus 58 percent of the production levies. The 2-percent levy set for A-type sugar beets raises the unit price of sugar to 46.72 euros per 100 kilograms, and the higher levy on B-type sugar beets raises its price to 32.42 euros per 100 kilograms (table 2).

Table 2. Components of the intervention price.

Item	Euros per 100 kilograms of: White sugar	Sugar beets
Intervention price, white sugar	63.19	
Transport cost of sugar beets	-4.41	
Processing cost	-24.36	
Value of molasses for beet growers	2.25	
White sugar price in beets (Price paid to beet growers)	36.67	4.767
Minimum beet price for A-sugar (98 percent of basic price)		4.662
Minimum beet price for B-sugar (60.5 percent of basic beet price)		2.884

Souce: USDA, ERS using data from Swedish Competition Authority (2002).

Components of the Intervention Price

The intervention price for white refined sugar is the price that the government agrees to pay to refineries on the basis that sugar beet processors will pay the growers the minimum beet price. In 2005, the intervention price was set at 63.19 euros per 100 kilograms. An equivalent intervention price is derived for raw sugar of a standard quality at 52.37 euros per 100 kilograms (83 percent of the white sugar intervention price) and is the price paid for raw cane sugar imported from the ACP countries. The sugar intervention price is increased for deficit regions (where consumption exceeds production); the premium equals the transport cost from the nearest surplus region. The sugar beet price is increased in a similar manner for deficit regions. The CMO for sugar also provides additional assistance to the sugar industry, namely "refining aid," which is granted to the sugarcane refining industry and covers certain costs of refining raw cane sugar.

For out-of-quota sugar (C-sugar), the price paid is determined by the returns to beet processors from sales of C-sugar on the world market. Although the CMO for sugar does not stipulate how much growers should be paid for C-sugar beets, intertrade agreements show that beet growers generally receive about 58 percent of exporting sugar processors' receipts for C-sugar. Usually, beet growers are paid for the C-sugar beets at the end of the marketing season, when all C-sugar has been exported.

Intervention purchases are the enabling mechanism backing intervention prices. However, intervention purchases are rare, as processors are able to export sugar with subsidies at prices typically above intervention prices. Only once in the last 25 years has sugar been offered to the intervention agencies: in marketing year 1986/87, the German Intervention Agency bought 15,703 tons of white sugar (Netherlands Economic Institute, 2000). In 2005, a substantial intervention store amounting to 850,000 tons resulted from countries overproducing in anticipation of the reforms. Moreover, the EU was forced to declassify to C-sugar nearly 1.9 million tons of quota sugar, which would have had to be exported with subsidies, thus exceeding the WTO limit.

Subsidies to Sugar-Using Industries

About 70 percent of total human sugar consumption in the EU comes from food and drinks, while the remainder is directly consumed. Sugar-using industries that benefit from production refunds under the CMO include the chemical and pharmaceutical industries, which receive compensation that effectively equalizes the prices that they pay to producers outside the EU. With production refunds, these industries (which represent less than 2 percent of total sugar consumption) are able to buy sugar at the equivalent of world market prices plus shipping costs. The production refund is fixed quarterly at a level equal to the average of the export refunds for a defined reference period minus 6.45 euros per 100 kilograms of white sugar equivalent. This deduction has the effect of denying the chemical industry sugar at world prices and forces the industry to use molasses or starch as its principal raw material. Food processors that use sugar also benefit from export refunds. Export refunds are fixed monthly and apply to sugar exported in processed products. This refund is fixed as a unit rate multiplied by the sugar content of the products. The sugar export refunds for processed products containing sugar operate in export transactions similar to the way that the EU's ingredient-based tariffs on processed foods and beverages containing sugar operate in import transactions.

EU Sugar and Trade Policies

The EU's external sugar trade is highly managed within the current CMO, principally through three instruments: "export refunds," contractual preferential agreements with third countries, and high import duties for nonpreferential sugar, including special safeguards. These policies regulate the two-way flow of sugar and have placed the EU in the unusual position of being both a significant importer and an exporter of sugar (Figure. 2).

Export Subsidies

Export subsidies form the key mechanism to dispose of surplus production and maintain high internal prices. Refunds are paid from the collected levies for all exported (A+B) sugar obtained from sugar beets or cane harvested in the EU and from sugar imported under the Protocol Agreement with India and the ACP countries. Export subsidies enable sugar producers/exporters to receive a guaranteed price for exported sugar, equivalent to the price they would receive in domestic markets. Export subsidies are intended to cover the difference between the EU price and the world price for sugar, allowing the sugar to be sold on the world market. Maximum export refunds are equal to the white sugar intervention price plus the shipping cost to the port (5 euros), minus the world price. The sum of export refunds plus the world market price gives producers nearly 670 euros per ton of sugar, inclusive of transport costs. Compared with the minimum price received in the EU internal market of 670 euros per ton (intervention plus transport costs), export refunds averaged 443 euros per ton for 2001/02, 485 euros per ton for 2002/03, and 480 euros per ton for 2003/04. Export subsidies can account for up to 75 percent of the EU intervention price. As shown in table 3, export refunds for EU sugar have remained fairly steady over the past decade.

Export subsidies are particularly critical to the large EU-based confectioners competing in world markets when one of the principal components of their exported product is sugar priced at roughly three times the world price. The EU has also used export subsidies to re-export 1 to 2 million tons of imported sugar under the Sugar Protocol with the ACP countries.

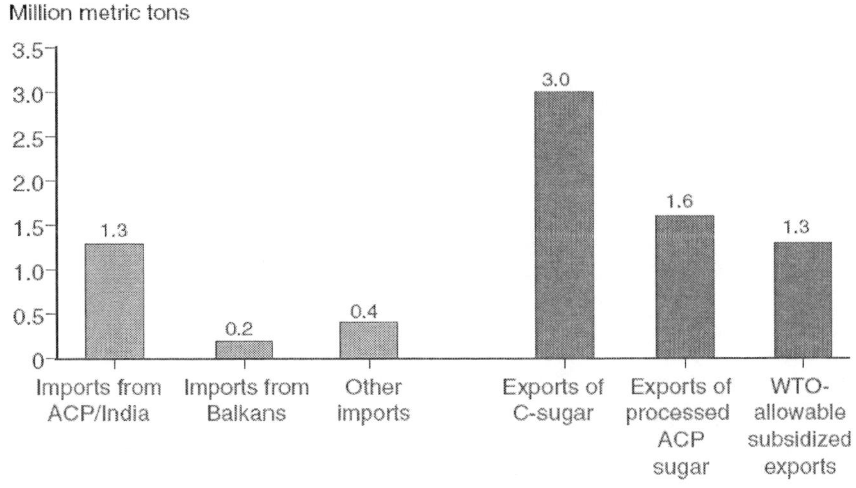

Source: USDA, ERS using data from AGRA Europe, EU Commission.

Figure 2. Structure of EU sugar trade, 2004.

Table 3. EU export refunds, 1996-2006.

Item	1996	1997	1998	1999	2000	2001	2002	2003	2004	2005	2006
					Million euros						
Refunds	1,230	1,360	1,370	1,593	1,497	1,008	1,168	1,021	1,285	1,081	1,117

Source: USDA, ERS.

Export subsidies are disbursed in two ways: weekly export tenders and standing export refunds. Export tenders are awarded each season. As part of the process, traders (and some processors) bid for the minimum level of refunds they can accept for exporting sugar, based on what they would receive for domestic sales. These tenders apply to most EU sugar exports. The standing export refund for sugar is meant to apply to the export of small quantities. Set every 2 weeks, the refund equals the lowest tender minus 3 euros per ton of sugar. Sugar-containing foods, drinks, and other products also qualify for export subsidies. These refunds are fixed monthly and are based on export refunds for white sugar minus 3 euros per 100 kilograms and the sugar content of the product.

EU-25 sugar exports, whether subsidized directly or indirectly, averaged about 6 million tons a year in the late 1990s and early 2000s and accounted for about 30 percent of total world exports. An EU average of 3.5 million tons of sugar (including sugar in processed products) is exported with a direct subsidy. This amount far exceeds the 1.273 million tons allowable under WTO commitments. The EU does not report to the WTO the subsidies given to re-exported refined sugar from ACP raw sugar imports as part of the required notifications. However, the recent WTO panel ruling clearly identified both refined sugar exports from imported raw ACP sugar and C-sugar exports as benefiting from export subsidies, making the EU noncompliant with its WTO obligations concerning sugar (World Trade Organization, 2005).

Export subsidies are paid for partly from producer levies (charged on A- and B-quota sugar) and partly from the EU budget (for re-exports of white sugar from imported raw ACP sugar). Overall, sugar export subsidies accounted for 75 percent of the total CMO budget for sugar in 2004 (table 4). This includes 802 million euros for the equivalent of 1.6 million tons of re-exported ACP sugar. By comparison, sugar production refunds for the chemical industry are 194 million euros (11 percent), refining aid for cane sugar totals 41 million euros (2 percent), export refunds for value-added food products containing sugar total 183 million euros (11 percent), and aid for the disposal of raw sugar (overseas department sugar) totals 18 million euros (1 percent).

Preferential Imports

Following the UK accession in 1973, the CMO for sugar incorporated some of the UK's earlier commitments to the ACP countries. This change led to the Sugar Protocol, which opened the EU market to a cane-sugar quota from 19 ACP countries. With the addition of a separate agreement that establishes an import quota of 10,000 tons of raw sugar for India, total imports under the Sugar Protocol amount to just over 1.3 million tons. These imports are exempt from duties, and the price paid to exporters is equal to the derived intervention price for raw sugar in the EU.

Table 4. CMO budget for sugar and export subsidy outlays, 2004.

	Million euros	Percent
Total budget for CMO for sugar	1,721	100
Total export subsidies (ES):	1,285	75
ES for ACP sugar	802	47
Refunds for chemical industry	194	11
Refining aid for cane sugar	41	2
ES for non-Annex I products	183	11
Aid for disposal for raw sugar	18	1

Source: USDA, ERS using data from CAP Monitor (2004).

In addition to imports under the Sugar Protocol, the EU imports sugar under preferential terms to meet its so-called Maximum Supply Needs (MSNs), based on the capacity needs of five member states for refining raw cane sugar.[5] Since 1995, MSNs were established for all member states, and extra sugar needs not met by production and imports under the Sugar Protocol are met by additional preferential sugar imports from India, the ACP countries, and regions within the French Overseas Department (DOM). This additional preferential sugar, totaling approximately 200,000 tons, can be imported duty free under bilateral agreements known as Special Preferential Sugar (SPS). Under the SPS agreements, exporters also receive the EU domestic price for raw cane sugar.

Since February 2001, the EU has agreed to offer progressive duty- and quota-free access to exports from 48 LDCs as part of the EBA Initiative. Access began with 74,185 tons in 2001 and will increase 15 percent annually until 2009, when free access will be allowed. The EBA Initiative ensures an increase in the flow of preferential imports into the EU, especially from low-cost countries, such as Zimbabwe, Ethiopia, Sudan, and Mozambique, as well as the possibility of implementing “SWAP” practices, which enable participating LDCs to import sugar from third countries at world prices and then export locally produced sugar to the EU. However, some in the EU question whether the EBA countries have the capacity to export significant amounts of sugar. Under the EBA Initiative, imports from eligible countries also have been counted as part of the SPS quota. EBA-eligible countries from the ACP group may choose between the EBA and Sugar Protocol agreements. Under either regime, suppliers receive the EU domestic price for raw cane sugar, but EBA countries will face no quota restrictions or duties after 2009.

In October 2007, the EU agreed to terminate the ACP Sugar Protocol by October 2009 to comply with the WTO panel ruling regarding sugar export subsidies, which includes the subsidies that went to re-export sugar from the ACP countries. The sugar arrangements between the EU and the ACP countries will depend on economic partnership agreements (EPA) that are negotiated between the EU and individual ACP countries or groups of ACP countries. The ACP countries will be granted duty-free access on January 1, 2010, for raw sugar and sugar beets but on the same terms as non-ACP developing countries. This development means that the “SWAPS” feature of the EBA Initiative, which allows EBA countries to import sugar at world prices and then export locally produced sugar to the EU, will not be available to the ACP countries that are not party to the EBA Initiative.

The EU also has a preferential import agreement with the Balkan countries, which can export within a tariff-rate quota of 193,000 tons to the EU duty free. Also, under the Uruguay

Round and as a result of the accession of Finland, Austria, and Sweden as member states, the EU agreed to a Most Favored Nation access quota covering 85,463 tons of raw cane sugar for refining, with reduced duties of 98 euros per ton for sugar from Cuba (58,969 tons) and Brazil (23,930 tons). This sugar does not qualify for a price guarantee. And, as mentioned earlier, the EU was obligated to allocate "current access" quotas of 490,000 tons to third countries to compensate for exports lost due to the accession of 10 Eastern European countries to the EU in 2004.

Import Duties

Import duties ensure that the price of imported sugar does not fall below the EU sugar price and that sugar imports from certain countries receive preferential status. It also provides for "special preferential" treatment of sugar imports from India and, starting in 2009, from the 48 LDCs as part of the EBA Initiative. Outside the preferential duties, border protection is very high and consists of two duties: one fixed and one variable resulting from the application of the special safeguard clause. The fixed duty is set at 419 euros per ton for white sugar and 339 euros per ton for raw sugar. The variable duty under the special safeguard clause is linked to the trigger price set at 531 euros per ton for white sugar and 418 euros per ton for raw sugar in the WTO Agreement on Agriculture. The EU can invoke the special safeguard clause whenever the world market price falls below the trigger price. Because the EU negotiated a high trigger price for sugar in the Uruguay Round, the safeguard clause and, hence, the additional duty have been applied consistently since 1995.

With the fixed duty and the special safeguard, the overall import tariffs on average amount to three times the world price. For example, if the world price is 250 euros per ton, the special safeguard will be 82.1 euros in addition to the fixed tariff of 419 euros, which results in a duty-inclusive import price of 751.1 euros per ton.[6] The high degree of protection from the combined fixed tariff plus special safeguard has kept nonpreferential imports at the minimum level. This effect can be illustrated through two measures. First, the EU's average overall duty was only 115 euros per ton in 2003 (87 euros per ton in 2002). Second, average annual imports of pure sugar at full import duty (nonpreferential) have been only 28,000 tons, compared with total imports of 1.5 million tons and the EU's annual consumption of 12.7 million tons. Thus, nonpreferential imports equal a mere 0.22 percent of consumption and 0.19 percent of the maximum EU sugar production quota.

Market Implications

Sugar beet production is regulated by national quotas and a high support system, making sugar beets one of the most remunerative crops in the EU. At 4.767 euros per 100 kilograms, the basic price for sugar beets is higher than the estimated breakeven price, at which point it becomes profitable to switch to alternative crops. Breakeven prices in the EU are estimated to range from 2.5 euros per 100 kilograms (Denmark) to 4.4 euros per 100 kilograms (Finland) (Commission of the European Communities, 2005). The distribution of production via national quotas favors the partitioning of national markets and ensures that sugar beets are grown across a wide range of conditions in terms of cost efficiency, from less efficient producers, such as Finland, Italy, Spain, and Greece, to more efficient producers, such as

France (with high beet yields), Belgium, Denmark, and the Netherlands (with efficient use of inputs) (see figure 1). Even within a particular region, the production quotas limit the competitive ability of efficient producers and create barriers to their entry to the market because sugar beet producers are tied to long-term supply contracts with regional sugar processors who hold sugar quota rights.

Supply response varies with price changes and depends on marginal cost. High-cost producers, supplying only within the A-quota, will continue to produce as long as the quota rent (defined as the difference between the price received and the marginal cost of production) is positive. Producers in some regions can supply quota (A and B) and out-of quota (C) sugar to the world market at world prices. For those producers, reductions in the intervention price (and hence the producer price) will cut into the captured quota rents but will not affect production under the existing price support regime (Frandsen et al., 2003). However, this scenario may not generalize to the overall EU supply response, particularly for producers whose marginal cost is on average higher than the world sugar price, which implies that a reduction in the intervention price would affect the level of EU sugar production and not just quota rents (Gohin and Bureau, 2006).

Domestic Market Price

In the EU sugar regime, intervention purchases have been rare. The principal mechanisms to manage supply and demand are production quotas, import controls, and the disposal of excess supply through subsidized exports. The end result of this multilayered support mechanism has been to create a price band within which domestic market prices fluctuate (Figure 3). The availability of intervention buying at intervention prices forms the floor, while full duty- paid prices for nonpreferential third-country imports form the ceiling. Under the CMO for sugar, subsidized prices can move over an approximate range from 670 euros per ton (which represents the export price plus applicable transport costs) to 750 euros per ton (depending on the world market price). The domestic EU sugar price has typically been maintained at or around three times the prevailing world market price.

Under perfect internal competition and given excess supply, sugar prices would be driven down to or near the price-floor level (intervention price plus transportation costs). However, prices rarely deviate from the price ceiling and are consistently kept far above the intervention price. The Netherlands Economic Institute (2000) examined price data on the sugar prices paid by industrial buyers in the EU and found that average prices in 1999 were 14 percent higher than the intervention price plus a storage levy (i.e., 743 euros per ton). Moreover, prices were high in both sugar-deficit and sugar-surplus regions, with no clear correlation between prices and supply balances. While investigating the merger between two sugar beet processors in 2001, the EU Commission found that the price level within the EU was 10- to 20-percent higher than the intervention price (i.e., ranging between 700 and 760 euros per ton). The market price for a large number of regions over several years was found to vary between 717 and 786 euros per ton, with the EU average at 746 euros per ton, according to the Swedish Competition Authority (2002) (Figure 4).

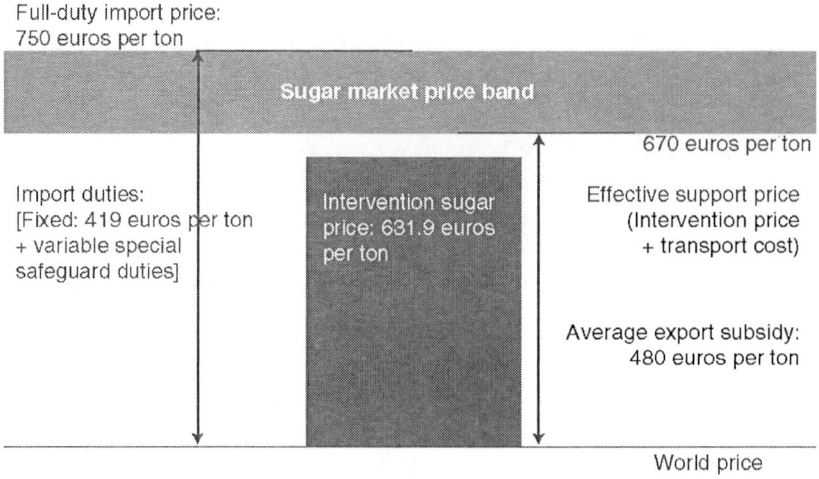

Source: USDA, ERS using data from Swinbank (2004).

Figure 3. EU sugar price support system.

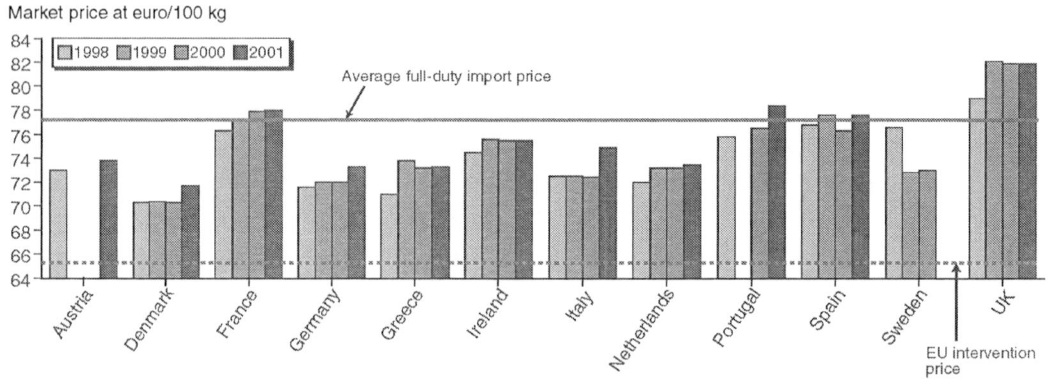

Source: USDA, ERS using data from AGRA Europe, EU Commission.

Figure 4. Market prices paid by sugar users by EU member states.

Sugar Industry Structure and Competition

Transport costs alone are not enough to account for the difference between observed sugar prices and expected prices under internal competition. Sweden's estimated average transportation cost is about 0.43 8 euros per 100 kilograms per 100 kilometers (400 kilometers is a typical distance between sugar plants in continental Europe), which is low compared with the price of approximately 70 euros per 100 kilograms (Swedish Competition Authority, 2002). An alternative explanation for the gap between market and intervention prices is the lack of competition in the sugar-processing industry and tacit collusion among firms.

Tacit collusion is a noncooperative type of collusion that does not involve any explicit communication among firms. Tacit collusion involves a restriction of output, coordination of prices, or market sharing (i.e., securing some parts of the market for certain firms) in which each firm independently or willingly engages. Each of these activities leads to an increase in price from which all firms benefit. Tacit collusion requires repeated interaction among firms, a price exceeding marginal cost that can be sustained by using the threat of retaliation in response to any attempt at competing to change the present market situation. If price is the variable that is coordinated, or if markets are shared, it is the threat of a future price war that keeps firms at the collusive equilibrium, where prices are substantially higher than marginal costs. A deviation from the collusive equilibrium is defined as a small unilateral decrease in price by one firm. A deviation lasts only as long as it goes unnoticed by other firms in the market. However, all firms are deterred from deviating by the threat of a price war during a retaliatory period, when all firms price at marginal cost and profits are low or zero for all firms. According to the Swedish Competition Authority (2002), firms in the EU sugar industry are able to charge higher prices through tacit collusion, and, thus, the exercise of collective market power without violating EU or national competition regulations.

The key factors necessary for tacit collusion to prevail are industry concentration, barriers to entry, significant transparency, and a retaliatory mechanism. In the EU-15, sugar processors are highly concentrated (see table 1). In 7 of 14 sugar-producing member states, only one firm holds the entire sugar production quota, while the remaining 7 sugar-producing member states have only a few large sugar beet processors. No clear pattern exists with respect to market structure that differs substantially across the member states. Even for countries with multiple sugar producers, such as France (15), Italy (5), and Belgium (3), prices are no lower than in countries with a single producer. The CMO for sugar has enhanced the ability of incumbent firms to sustain tacit collusion through higher market concentration. With fewer firms, it is easier for firms to determine who has deviated in order to retaliate. The CMO for sugar has maintained production in regions that otherwise would not cultivate sugar beets. By assigning quotas on a national level, the CMO for sugar has consolidated the division of national markets. Without the CMO's quota system, it would be possible to decrease concentration on the national level.

In the EU sugar market, new entry of firms is restricted because a production quota is necessary, and quotas are allocated only to incumbent firms. The tariff structure makes it difficult to import sugar from nonpreferential countries, hence preventing competition from outside the EU. Also, the CMO for sugar has limited the possible entry of firms producing the sugar substitute isoglucose via restrictive production quotas.

Substantial transparency in the market facilitates tacit collusion. In the sugar industry, transparency refers to a firm's ability to detect whether a competing firm has deviated by offering better deals to sugar buyers (without knowing the price agreed between a supplier and a buyer). The CMO for sugar has increased transparency by separating markets geographically and fostering market sharing, thereby reducing the number of alternative suppliers.

For tacit collusion to be sustainable, threats of retaliation need to be credible (i.e., there must be some effective retaliatory mechanism that yields lower profits for all firms in the case of deviation and retaliation). In the EU sugar market, subsidized annual exports of A- and B-sugar, which average 2.9 million tons (compared with domestic sugar consumption of 12.7 million tons), provide an opportunity for retaliation by competing firms. As retaliation to an observed deviation by a firm, other firms could redirect some of their sugar normally intended for exports into the domestic market, hence starting a price war by bringing prices down to the marginal cost level (the intervention price plus transport cost). Consequently, firms are deterred from deviating by the threat of a price war during a retaliatory period, where all firms are forced to operate at marginal cost and zero profits. By assigning quotas in excess of consumption, the CMO for sugar has in effect provided an effective retaliatory mechanism.

World Sugar Markets and Prices

The EU plays a significant role in the world sugar market, accounting for 12 percent of world exports and 5 percent of world imports in 2004. On average, international sugar trade amounts to roughly 40 million tons, or 30 percent of world production. Although the leading sugar-producing countries are also the leading consumers, sugar is a widely traded commodity. Among the main features of world sugar trade are the differentiation between raw and refined sugar, the regional character of trade flows, and the existence of policies and subsidies that affect a significant portion of the sugar trade.

The differentiated nature of traded sugar derives from its two main sources: sugarcane and sugar beets. Cane sugar constitutes about 70 percent of world production (30 percent for beet sugar). Sugarcane processing has two stages. The first stage yields semiprocessed (or raw) sugar that can be transported and exported. Much of the sugar imported by the EU is in this form. Once imported, the raw sugar undergoes a second processing stage to produce a white crystal sugar (refined). Sugar beets, on the other hand, are processed domestically in a single step into crystal sugar. The EU produces mostly beet sugar, processed domestically and supplied to either domestic or international markets. The EU also re-exports excess refined sugar derived from imported raw cane sugar.

World sugar trade is also characterized by regional patterns (table 5). Australia's sugar exports to the EU-25 in 2000-02 averaged $8 million, a small amount of the country's total sugar exports of $799 million. The main raw/refined sugar exporters are Brazil, the EU, Central America, African producers, Thailand, and Fiji.

Table 5. Bilateral trade flow: All sugar, 2000-02.

	Importer							
Exporter	EU-25	Japan	U.S.	Russia	N. Afr.	M. East	Other	Total
				$ millions				
ACP	661	37	147	325	33	129	496	1,828
EU-25	0	27	18	88	327	448	788	1,696
Brazil	11	0	76	299	139	269	477	1,271
Australia	8	126	38	0	14	52	561	799
Thailand	1	121	5	42	0	36	452	657
Colombia	0	0	25	46	3	0	152	226
LDC	95	0	28	0	10	3	85	221
India	26	0	1	0	0	7	142	176
Other	382	131	476	124	65	153	1,334	2,665
Total	1,184	442	814	924	591	1,097	4,487	9,539

Note: LDC=Least developed countries. Totals reflect average over 2000-02.

Source: USDA, ERS using data from Dimaranan and MacDougall (2006) (average 2000-02).

Table 6a. Top 20 raw sugar exporters to the world and the EU-15, 2003.

Raw sugar exports				
Exporter	**To world**	**Exporter**	**To EU-15**	**Share of world**
	$ millions		*$ millions*	*Percent*
Brazil	1,350	Mauritius	298	99
Thailand	426	CARICOM	153	76
Cuba	364	Fiji	105	88
Mauritius	302	Guyana	90	72
Guatemala	212	Malawi	44	47
CARICOM	201	Cuba	33	9
South Africa	151	Belize	23	63
Guyana	125	Barbados	21	100
Fiji	120	Brazil	19	1
Colombia	107	Trinidad and Tobago	13	100
Malawi	92	Aruba	13	100
India	69	Zambia	12	39
Philippines	62	Ethiopia	10	63
El Salvador	47	Sudan	7	84
Belize	36	St. Kitts and Nevis	7	99
Zambia	31	Kenya	6	31
Costa Rica	25	Cote d'Ivoire	6	89
Barbados	21	Nepal	5	95
Kenya	20	Paraguay	3	28
Peru	19	United States	3	43
Argentina	17	Madagascar	1	99
EU-15	90	EU-15 (intra-trade)	85	94

Note: CARICOM member states include Antigua and Barbuda, the Bahamas, Barbados, Belize, Dominica, Grenada, Guyana, Haiti, Jamaica, Montserrat, St. Kitts and Nevis, Saint Lucia, St. Vincent and the Grenadines, Suriname, and Trinidad and Tobago.

Most of the EU's raw sugar imports come from African, Caribbean, and Central American countries and Fiji under the Sugar Protocol (table 6a). Brazil exports raw sugar mostly to Russia, North Africa, the Middle East, and North America. For refined sugar, the top three exporters are the EU, Brazil, and Thailand (table 6b). Most of the EU's refined sugar exports go to other parts of Europe, the Middle East, North Africa, and, to a lesser extent, Africa and, much less, Asia. Brazil also exports heavily to the Middle East and Africa. However, neither the EU nor Brazil dominates in Asian markets, which source most of their sugar needs regionally, notably from Thailand and Australia.

Sixty percent of world sugar trade takes place under long-term contracts, preferential agreements, and subsidies, while only 40 percent trades at world market prices. Within this global sugar situation characterized by residual demand, differentiated raw/refined sugar markets, and regional trade patterns, the EU plays a significant role in influencing world prices, particularly for white sugar, mostly due to large quantities of subsidy-enabled sugar exports. Of the 21.3 million tons of white sugar exported worldwide, the EU exports over 6.4

million tons, a large share of which benefits from export subsidies, which tend to put downward pressures on prices.

Protectionist policies and subsidies also contribute to the volatility of world sugar prices (Figure 5). For example, raw sugar prices tripled from 1985 to 1989 and more than doubled between 2003 and 2006, compared with much lower price variations for wheat. Given the high degree of intervention in the sugar market, international prices tend to be extremely volatile, following an erratic path. Factors accounting for the inherent volatility of the world price of sugar include the perennial nature of sugarcane production, the high cost of switching to alternative crops, and limited possible options for sugar beet producers due to complex capital investments, transport, and processing. Such investments require stable output even when prices are low, hence, exacerbating excess supply. Variations in exchange rates, which may mask world price signals, coupled with high levels of government support and trade protection, also contribute to world market prices that are sometimes even lower than the marginal cost of the most efficient producers (OECD, 2005).

Since 1995, world sugar prices have trended downward, mainly due to an excess of production over consumption. During the 1990s, monthly world prices for raw sugar fluctuated between 280 euros per ton (March 1990) and 110 euros per ton (April 1999). From their low point in 1999/2000, prices improved over 2000/01, reaching an average of 240 euros per ton. By the following year, prices had declined again to 180 euros per ton. Since 2003, however, sugar prices have increased, largely driven by high oil prices and the sharp increase in ethanol-related sugar demand. The strong co-movement between rising crude oil and sugar prices also is tied to the close link between ethanol and sugar production in Brazil, the world's largest sugar producer (20 percent of the world's total) and exporter (38 percent of the world's total). Any increase in gasoline prices stimulates Brazilian domestic demand for ethanol, reduces sugar exports, and raises sugar prices. In the short run, at least, world sugar prices likely will be tied to developments in the ethanol market.

Source: USDA, ERS (2006).

Figure 5. World sugar and commodity prices Dollars per ton.

Table 6b. Refined (white) sugar trade, 2003.

Refined sugar exports				
Exporter	**To world**	**Exporter**	**To EU-15**	
	$ millions		*$ millions*	
EU-15	2836.1	EU-15	1738.7	
EC-10	274.2	EC-10	152.1	
Brazil	800.4	Pakistan	43.1	
Thailand	549.2	United States	20.0	
India	195.6	Egypt	18.4	
Belarus	132.4	Cote d'Ivoire	14.8	
Colombia	127.6	Canada	13.1	
Canada	126.7	Brazil	10.8	
Malaysia	96.7	India	9.8	
Pakistan	74.3	Tanzania	8.4	
United States	74.0	Sudan	6.5	
South Africa	73.6	Switzerland	4.8	
South Korea	71.9	Thailand	4.3	
Turkey	47.4	Mexico	3.2	
Kazakhstan	37.6	South Africa	3.0	
Australia	34.1	Morocco	2.2	
Georgia	34.0	Macedonia	2.1	
Egypt	33.1	Mauritius	1.9	
China	26.7	CARICOM	1.7	
Singapore	24.1	Malawi	1.5	
Refined sugar imports				
Importer	**From world**	**Importer**	**From EU-15**	**Share of world**
	$ millions		*$ millions*	*Percent*
EU-15	2,355.9	EU-15 (Intra-trade)	1,838.2	
EC-10	123.1	EC-10	89.8	
United States	234.3	Israel	91.2	98
Syria	150.3	Algeria	90.6	76
Indonesia	150.3	Switzerland	65.2	97
Nigeria	139.0	Norway	53.6	89
Algeria	119.1	Lebanon	40.3	97
Sri Lanka	108.6	Nigeria	39.8	29
Israel	93.0	Syria	38.7	26
Singapore	91.1	Indonesia	33.5	22
Ghana	84.3	Sri Lanka	30.7	28
Switzerland	67.2	Tunisia	30.0	65
Norway	60.6	Jordan	20.4	41
Mexico	54.0	Armenia	16.4	75
Egypt	50.7	Senegal	13.5	85
Russia	49.9	Croatia	13.1	87
Jordan	49.4	Macedonia	10.7	84
South Korea	48.4	Albania	10.1	66
Chile	46.8	Gambia	9.9	57
Tunisia	46.2	Romania	9.0	30

Source: USDA, ERS using data from COMTRADE.

Table 7. Key elements of the EU's 2005 sugar reform.

Item	Previous policy regime	Policy change under reform
Production quotas	National allocation of quotas; A and B quota (17.441 MT); (industrial sugar excluded)	Merging A+B; quota buyout over 4 years
Overquota sugar ("C-sugar")	Solely for exports; subject to declassification/carryover	C-sugar cannot be exported; penalty with super levy
Quotas for sugar-substitute	Isoglucose (500,000 tons), inulin (300,000 tons); no intervention	Isoglucose quota increase by 403,000 tons
Intervention price	White sugar: 631.9 euros per ton; minimum sugar beet price 46.72 euros per 100 kilograms (A); 32.42 euros (B)	Intervention becomes reference price for triggering private storage aid; 36 percent cut over 4 yrs
Intervention purchases	In place; rarely used (1987, 2005)	Phased out in 4 years
Excess supply management	Private storage subsidies	Private storage/carry forward to next year quota
Farm payments	None	64 percent of price cut; included in Single Farm Payment (new CAP)
Set-aside	None	Set-aside + energy use subsidies
Export subsidies	A+B sugar; re-export processed ACP sugar	Reducing export subsidies to WTO limits—Via lower production; no exports of C-sugar
Preferential imports	ACP/India "Sugar Protocol"; Special Preferential Sugar; EBA Initiative; Balkan TRQ; MFN quota access	No change
Duties/Special safeguards	High fixed overquota tariff/ Special safeguard duty	No change

The 2005 Sugar Reform

The EU Commission issued a draft proposal for sugar reform in July 2004 that was subsequently revised following the WTO ruling on EU export subsidies. A revised reform package was proposed in June 2005, endorsed by the Council of Ministers in November 2005, and formalized by the EU Parliament (January 19, 2006) and later by the Council of Ministers (February 22, 2006). The stated aims of the reform are (1) to encourage reductions in domestic sugar output, particularly in regions with high production costs or lower sugar beet yields; (2) to bring export subsidies in line with WTO commitments; (3) to dampen incentives for EU sugar imports from the EBA countries; and (4) to reduce the price gap between sugar and competing sweeteners to forestall the substitution of sugar.

The new elements of the new sugar regime, which took effect on July 1, 2006, include lower price support, a rationalization of the production quota system through a voluntary buyout scheme, and compensation for farmers paid through the Single Farm Payment (SFP), a fixed producer payment based on historical aid entitlements, which is a key feature of the new CAP regime. The main elements of the reform, summarized in table 7, are as follows:

- The intervention price for sugar is cut 36 percent (from 631.9 euros to 404.4 euros per ton) over 4 years, beginning in 2006/07 (minimum sugar beet price is reduced 39.5 percent to 263 euros per ton).
- In exchange for cuts in support prices, farmers will receive direct compensation at an average of 64.2 percent of the price cut. The aid is included in the SFP (and, hence, partly decoupled from production) and is linked to compliance with a set of environmental and land management practices. Farmers in those countries giving up at least 50 percent of their quota receive an additional coupled payment (i.e., linked to production) of 30 percent of the income loss for a maximum of 5 years, plus possible limited national aid. In addition, sugar beets qualify for set-aside payments when grown as a nonfood crop (i.e., sugar beets for nonfood use may be grown on set-aside area) or may receive the energy crop aid of 45 euros per hectare.
- The current quota system is rearranged by merging A and B quota into a single production quota totaling 17.4 million tons for the EU-25. To maintain a certain level of production in the current C-sugar-producing countries, an additional amount of 1 million tons will be made available as new quota (plus 100,000 tons for other countries) to be purchased with a one-off payment of 730 euros per ton. Moreover, a production charge of 12 euros per ton is imposed on producers for all quota production. Production quotas exclude sugar for chemical, pharmaceutical, and ethanol production.
- To encourage the buyout of quotas, a voluntary restructuring scheme is in place for 4 years and covers EU sugar factories and isoglucose and inulin syrup producers. The scheme involves buying out quota from sugar processors and encouraging factory closures. The payment is 730 euros per ton in years 1 and 2, falling to 625 euros per ton in year 3 and 520 euros per ton in the final year. An additional diversification fund is set for member states where quota retirement is larger than expected. Both payments will be financed by a levy on quota holders lasting 3 years. The first-year levy is equal to 126.4 euros per ton; the second-year levy, 173.8 euros per ton; and the third-year levy, 113.0 euros per ton. The isoglucose levy is fixed at 50 percent of these rates. To discourage production outside quota (C-sugar), a prohibitive "super levy" is applied to over- quota production.
- Even though the quota release will be handled under a voluntary scheme, the production quota for 2006/07 was temporarily cut by 2 million tons, or 11 percent, to prevent overproduction.
- Supply management tools include both old and new mechanisms. A key change is the 4-year phaseout of the intervention system, which is replaced with a reference price. Intervention purchases will not be available after the transition period. Instead, supply management will rely mostly on a private storage system when the market price falls below the reference price. Other available supply management tools

include the option to carry forward quota overrun to the following marketing year, temporary quota withdrawal, or use of excess production for industrial products or for supplying the outermost regions of the EU. However, no exports of declassified sugar (C-sugar) are permitted under the new regime.

- For nonsugar sweeteners, a quota increase for isoglucose of 403,000 tons for the existing producer companies is being phased in over 3 years. This measure includes the option to purchase extra isoglucose quota in Italy (60,000 tons), Sweden (35,000 tons), and Lithuania (8,000 tons) at the restructuring aid price set in the first year at 730 euros per ton.
- The border protection regime did not change, aside from reducing exports and subsidies to WTO limits and maintaining the existing international import commitments. Nonpreferential import duties (including the special safeguards) are not affected by the new regime. Only a new WTO agreement from the current Doha Round of negotiations would affect these import policy instruments.
- In September 2007 (AgraEurope, 2007), the EU agreed to changes in the sugar-restructuring plan to encourage removal of EU sugar quota by offering higher buyback payments to beet growers while shielding refiners from receiving any less than 90 percent of the buyback payment. However, growers can apply directly to the Commission for 237.5 euros per ton payment as long as the renounced quota does not exceed 10 percent of the refiner's total quota. Also, factories that give up quota in 2008/09 will be reimbursed for the restructuring levy they paid in 2007/08. By the end of 2007, sugar producers had renounced 1.56 million metric tones of quota, more than twice that of the previous year.

What then are the likely impacts of the EU sugar reform on supply, demand, and prices in the EU? How are sugar imports from both preferential and nonpreferential sources likely to change? And, what are the likely effects on world sugar prices, if any, that would arise from the reforms?

Reform Assessment

An examination of the current market reveals some indications of the potential effects of the EU sugar reforms. While the new regime represents a significant departure from existing policies, many of the support instruments are still in place, which has the potential to limit the extent to which the goals of the reform can be achieved. Even with cuts in support prices and incentives to renounce quota, sugar crops may still be more attractive to growers than the alternatives. Moreover, high import barriers will continue to shield the domestic sugar industry. The inability of quota holders to trade quotas across member states may restrict the degree of industry adjustment toward greater cost efficiency. In addition, the new policy continues to limit substitutions between sugar and alternative sweeteners, as the new isoglucose production quotas are still highly restrictive.

The new provisions of the EU sugar regime are expected to alter the EU sugar market. According to EU Commission estimates, restricting EU sugar exports to comply with the WTO panel ruling alone will require a reduction in EU sugar production of around 2 million tons per year. High-cost regions such as Greece, Ireland, Italy, and Portugal would take the brunt of the reduction, while low-cost regions could increase production, depending on the restructuring component of the proposal. France, Germany, the Netherlands, Belgium, and the UK have been the largest producers of over-quota sugar. Nonetheless, regions within these countries are likely to see sugar production decline because of high production costs and/or inefficient processors.

Regarding trade, while the EBA countries are expected to increase sugar shipments to the EU after 2009, there is some uncertainty about the expected volume of imports from these countries. Because of the "SWAPS" provision in the EBA Initiative, EBA countries would be able to import sugar at world prices and then export locally produced sugar to the EU. Even with lower intervention prices under the reform, some in the EU sugar industry doubt that the EBA countries will be able to export raw sugar to the EU profitably at the lower intervention prices. Under the reform, ACP countries within the LDC group (i.e., Malawi, Zambia, and Zimbabwe) have the potential to offset losses in their current quota exports to the EU by increasing their export volumes under the EBA Initiative (Chaplin and Matthews, 2006).

Outside the EBA countries, the reduction of intervention prices will have adverse effects on the ACP countries, which enjoy preferential access to the EU's high-priced sugar market. Moreover, any gains by preference-receiving countries due to higher exports under higher world prices (due to the EU reform) will depend on the proportion of their exports to non-EU countries and the extent to which their exports benefit from preferential agreements with other countries. Among non-LDC ACP countries, the adverse effects would be greater for high-cost producers that depend heavily on the EU for their sugar exports, such as Mauritius, Fiji, and Guyana. These countries, having had secure market access to the EU, lacked the incentives to improve their competitiveness, leading to high production cost structures over time (Larson and Borrell, 2001). In October 2007, the EU scrapped the ACP Sugar Protocol (AgraEurope, 2008) as a part of the economic partnership agreements (EPA) that would inter alia reduce tariffs on raw sugar to zero by the end of 2009.

The trade implications of the EU reform are ambiguous outside preferential trade. On the import side, little change is likely given the unchanged high import barriers (outside preferential access). The EU sugar reform, therefore, may have a limited effect on competitive sugar suppliers currently excluded from the EU market. The implications of the reform may be felt more via effects on world prices. A key determinant is the extent of the sugar price rise, as the EU withdraws a significant share of refined sugar exports from the world market. Higher prices for refined sugar may adversely affect importing regions but may benefit competing sugar exporters. Such price effects are unlikely to be neutralized by raw sugar exporters, such as Brazil, as it faces greater demand for sugar by domestic ethanol producers, further contributing to higher world sugar prices.

The reform is expected to alter EU sugar production significantly. However, the key issue of whether domestic sugar prices will follow cuts in intervention prices hinges largely on whether the reforms can weaken the industry's ability to engage in tacit collusion in sugar pricing. The tendency toward the consolidation of processors is likely to reinforce market power within the industry and strengthen oligopolistic and tacitly collusive behavior. As such, market prices may not fall as much as the scheduled reduction of intervention prices might imply. A decrease in the intervention price alone would have no effect on concentration, transparency and entry into sugar processing, or import barriers.

A reduction of export subsidies in compliance with WTO constraints would also make collusion easier to sustain. Under this scenario, low-cost firms still can export at a profit (through lower quantities as profits are reduced from lower subsidies), while high-cost firms reduce their output by the amount previously exported. Because the majority of the EU's sugar-producing firms are likely to belong to the low-cost category, a reduction in export subsidies will have the effect of lowering output (from exiting high-cost firms) and may, in fact, increase the incentives for tacit collusion. With lower profits from exports, low-cost firms have a greater incentive not to deviate from tacit collusion because the profit loss under retaliation from domestic sales is now proportionally higher (Swedish Competition Authority, 2002).

The effect of a quota reduction on tacit collusion is indeterminate, particularly if the reduction is proportional to all firms. Since it is more profitable for firms to sell sugar to the domestic market than to export, the quota reduction reduces the amount of sugar available for exports. Hence, both the ability to deviate and the ability of competing firms to retaliate (by shifting export quantities to domestic sales) are weakened. If the quota reduction results in fewer quota holders within the EU, concentration will increase, and as a result, tacit collusion may be facilitated.

The effect of quota tradability on tacit collusion depends on the extent of firm concentration. If the tradability of quotas among regions (or states) leads to increased firm concentration, then the effect on tacit collusion will be indeterminate because the increase in collusive profits will counteract the lessened ability to divide markets among firms. If, however, quota tradability reduces concentration through new firm entry, then tacit collusion will be reduced, through both the reduction of collusive profits and the lessening of the ability to divide markets among firms.

Certain policy changes not included in the new reform could have had the effect of weakening the industry's collusive behavior and, more significantly, altering the economics of EU sugar. For example, lower import tariffs would reduce both the incentives for tacit collusion and the market price, and introduction of the ability to trade quotas across member states would make tacit collusion harder to sustain. Furthermore, unrestricted production of HFCS (isoglucose) would make collusion difficult to sustain in markets where HFCS is a substitute. However, as these policy instruments are not part of the sugar reform, the net effect of the reform on the industry's competitive behavior and producer incentives is unclear.

A Multi-Model Economic Analysis

A properly specified quantitative model of the EU sugar regime must reflect the regime's key features: production quotas, price support, export subsidies, and import policies, including preferential import agreements. No single model can adequately capture all of the subtleties of the EU sugar regime. This analysis applied three economic models and focused on their distinctive features to provide a richer assessment of the EU sugar reform (see box, "Descriptions of the Models"). A partial equilibrium (PE) model called the Partial Equilibrium Model of Agricultural Trade (PEATSIM), which captures in detail many policy instruments of the EU's sugar sector and the CAP, projected likely changes in EU production, consumption, net trade, and prices. A second PE model called the U.S. Sugar and Sweetener

Model, which allows for substitution between sugar and alternative sweeteners, generated detailed analysis of the U.S., Mexican, and EU sweetener industries. Finally, a special version of the global computable general equilibrium (CGE) model created by the Global Trade Analysis Project (GTAP) was used to account explicitly for bilateral trade, tariff rate quotas, export subsidies, and imperfect competition in the sugar industry. This multimodel approach enabled a broader examination of the EU sugar reform than could be obtained from any of these models individually.

Partial Equilibrium Analysis

The PEATSIM model was used to compare the EU sugar "reform scenario" with an alternative "status quo scenario" of no reform. Under the reform scenario, the intervention price is cut by 36 percent, the production quota is cut by 15 percent, and exports are cut by 53 percent to represent the removal of C-sugar exports. Under the status quo scenario, the expected impact of EBA imports induces lower domestic market prices, resulting in a downward adjustment of the producer price by 11 percent. The model baseline under both scenarios was run from 2004 to 2009. Results are shown in table 8.

Under the reform scenario, model results show that the combined effect of cuts in the intervention price and production quotas will yield a 17.3-percent contraction of EU sugar output by 2009 (compared with the status quo scenario), while consumption will rise by 7.68 percent over the 5-year period (compared with the status quo), or 1.53 percent annually. Consumer prices will fall 28.8 percent, compared with the status quo (see table 8, difference between third and fifth column). Under the reform scenario, the supply- demand balance will be reversed, with production declining and consumption increasing such that domestic consumption exceeds production. Since exports are exogenously cut relative to the base year to reflect the removal of C-sugar exports, imports will need to expand to 5.1 million tons (compared with the status quo), or 113.5 percent, to balance available supplies and demand. However, as the model computes only total country trade, these imports are not separated between EBA and non-EBA sources.

Results from the U.S. Sugar and Sweetener Model feature a larger production response (-34.3 percent by 2010) under the reform scenario, compared with that of the status quo case.[7] This higher level of output contraction is due to the irreversible loss of production capacity once prices fall below variable costs.[8] The analysis assumes that EBA sugar exports to the EU will grow to 1.5 million tons by 2015, the end of the projection period. However, as the model calculates net trade balances, import flows are not separated between preferential and nonpreferential sources. Despite an 11.6-percent decrease in the price (compared with the status quo case), EU sugar consumption will increase modestly (by about 470,000 tons, or about 2.8 percent, over the period 2007-10). EU-25 sugar exports fall to insignificant levels by 2010 and remain at or below WTO commitment levels through the end of the projection period. Similar to the PEATSIM results, the U.S. Sugar and Sweetener Model results show that imports will increase 91 percent by 2010 to equalize prices across the EU-15 and the new member states.

DESCRIPTION OF THE MODELS

PEATSIM

PEATSIM is a PE, multicommodity, multiregion agricultural trade model (Stout and Abler, 2003). With a net trade specification, the model sets a region to be either a net importer or a net exporter. In this case, ERS treats the EU as a net sugar importer since researchers are most interested in import responses to the EU sugar reform and since exports are set exogenously to be reduced as part of a policy change. Sugar production in the model is handled on a raw sugar equivalent basis, and since preferential agreements are not explicitly addressed, production is adjusted to include preferential imports (from the ACP countries, India, and the Balkans). The model specifies a support price in the form of a reference price treated as an intervention price. For the import regime, import tariffs are based on two-tier tariffs using the ad valorem equivalent of specific tariffs. Under the EU policy reform, sugar farm payments equivalent to 64.2 percent of price cuts are introduced into the model starting in 2006 and treated as decoupled payments. This means that the payments are not expected to affect production decisions. Following the 2003 CAP reform, set-aside payments are ignored, given that their reallocation is linked to all arable crops (and not just grains and oilseeds, as was the case before the 2003 CAP reform). Therefore, one would expect little influence on production patterns.

For supply management, the model is initially set so that the quota is equal to beginning production. To model the impact of the reforms besides cutting the intervention price by 36 percent, the future target production quota is set to balance consumption, preferential imports, and WTO-permitted exports. For the production quota, a phased quota reduction is implemented in the reform scenario from 2006 to 2010. The desired quota level in 2010 is determined by taking into account the EU's WTO export commitment, preferential imports, and domestic consumption. Therefore, in the reform scenario, a target production quota cut is induced in line with the EU Commission's decision to cut quotas by 15 percent for 2006/07, which aims to achieve the supply-demand balance. Moreover, the model scenario "restrains" exports by preventing C-sugar from being exported, thus implicitly accounting for EU compliance with the WTO ruling on export subsidies.

U.S. Sugar and Sweetener Model

The U.S. Sugar and Sweetener Model is a PE model with greater detail on sugar and sweetener production, supply, and demand (Haley, 2006). The model is specified for the United States and Mexico and extended to the EU, with the rest of the world as a residual. The model is initialized to a 2003 base and generates yearly baseline projections to 2015. The model covers sugar, substitute sweeteners (high fructose corn syrup or high fructose syrup), and the primary sugar crops (cane sugar and sugar beets). Effects of policy changes, such as the EU sugar reform, can be compared with the model's baseline.

The EU component of the model includes the intervention/reference price and sugar and isoglucose production quotas. Because the model does not permit bilateral trade flows, preferential imports (from ACP, EBA, and Balkan regions) are treated as a composite, while over-quota sugar imports are subject to over-quota tariffs. The EU sugar and isoglucose supply response to policy changes is modeled via a logistic function that determines changes in processing capacity and, hence, production. Area planted with sugar beets has been stable because EU sugar policies limit the linkages between the world price and domestic production and given the inelastic response of price changes to planted areas and consumption. Policy-induced production change is modeled as follows. Based on estimates of sugar and isoglucose variable costs, the simulation-induced drop in producer returns (to or below the variable cost level) leads to an irreversible reduction of processing capacity by 50 percent relative to the base period. As processing capacity exits, sugar beet area devoted to that capacity exits as well.

GTAP

The applied general equilibrium model GTAP (Hertel, 1997) is a multiregion, multisector global model of production and trade that follows the standard theoretical specifications of CGE trade models. This analysis uses a special version of GTAP that deals with imperfect competition and the oligopolistic behavior of firms (Francois, 1998; Elbehri and Hertel, 2006). The model distinguishes between the sugar industry and the sugar beet/sugar cane crop sectors. Given the market structure of the sugar industry and its collusive practices, sugar processing is modeled under oligopolistic competitive behavior with increasing returns to scale. This allows the model to separate cases with or without firm entry or exit and, hence, enables separate assessments of production adjustments at the firm and industry levels.

In this analysis, only the sugar industry is treated as oligopolistic with scale economies, while other sectors are assumed to be perfectly competitive. Calibration of the oligopoly model and scale economies for the sugar industry was based on sugar manufacturing data using firm-level sugar production and deriving the Herfindhal index for all EU countries that produce sugar. The maximum-derived Herfindhal index (estimated for Portugal) was also applied to the remaining member states that operate a single plant (firm). To calibrate the Cournot model, Herfindhal indices and the model-conformable benchmark number of firms were combined with the perceived demand elasticity for each sector. (For a detailed description of GTAP's oligopoly speci-fication, see Elbehri and Hertel, 2006.) The benchmark price-cost margins (markups) are derived for the imperfectly competitive sectors using GTAP trade shares and the model's elasticities of substitution. These markups are consistent with positive profits in the benchmark data set. Model implementation also requires the calibration of a cost disadvantage ratio (CDR) reflecting unexploited economies of scale. The CDRs for the EU were taken from Harrison, Rutherford, and Tarr (1994). Data for non-EU regions are reported by Roland-Holst, Reinert, and Shiells (1994). The scaled ratios are reported in table 4 and range from 0.016 for the EU to 0.001 for non-EU regions.

Sugar production in the EU is also constrained by quota limits. The sugar production quota is modeled under a complementarity condition allowing for endogenous regime change from binding to nonbinding with associated quota rents (under binding quotas) expressed as a tax equivalent. The model does not separate between quota (A and B) and nonquota (C) sugar. Rather, the model takes into consideration the share of C-sugar in total sugar production in setting the initial degree of quota binding status. Except for Italy, all 13 EU regions start from a binding quota position.

Sugar is treated on a raw sugar equivalent basis, and the sugar import regime is modeled as a system of bilaterally allocated import quotas following the Armington specification. The model includes 28 regions, 14 of which are EU regions (Belgium, Denmark, Germany, Spain, France, Italy, the Netherlands, Austria, Poland, Sweden, the UK, Hungary, Czech Republic, other EU) and 14 non-EU regions (United States, Canada, Mexico, Brazil, China, India, Australia, Thailand, ACP, LDC, Mediterranean, Central America, other Asia, and rest of world). Preferential imports are treated via bilateral tariff-rate quotas for a specific set of partners (ACP, India). Moreover, these preferential exporters are assumed to capture the quota rents to mimic the fact that these countries receive guaranteed prices equivalent to domestic prices.

Table 8. Comparing EU reform to status quo: EU production, consumption, trade, and trade effects.

Item	Base year*	Reform scenario	Change from base	Status quo scenario	Change from base
	PEATSIM results (2009)				
	Million tons		*Percent*	*Million tons*	*Percent*
Production	20.6	17.0	-17.3	20.6	0.0
Consumption	18.4	20.0	8.9	18.6	1.2
Exports	4.4	2.0	-53.1	4.3	-1.5
Imports	2.4	5.1	113.5	2.4	-0.7
	Dollars per kg			*Dollars per kg*	
Producer price	0.6	0.4	-39.2	0.6	-10.6
Consumer price	0.6	0.4	-39.3	0.6	-10.5
Export price	0.6	0.4	-39.3	0.6	-10.5
Import price	0.6	0.4	-39.3	0.6	-10.5
World price	0.2	0.2	8.0	0.2	-1.5
	U.S. Sugar and Sweetener Model results (2010)				
	Million tons		*Percent*	*Million tons*	*Percent*
Production	22.2	14.6	-34.3	21.6	-2.7
Consumption	16.1	16.7	3.7	16.2	0.9
Isoglucose consumption	0.6	0.7	7.8	0.5	-24.9
Exports	6.0	0.2	-97.0	8.1	36.4
Imports	2.1	4.0	91.9	2.9	37.1
	Dollars per kg			*Dollars per kg*	
Producer price	0.7	0.4	-37.7	0.5	-26.1
World price	0.2	0.2	15.0	0.2	-15.0

* Base period is 2004 for PEATSIM and 2003 for the U.S. Sugar and Sweetener Model.
Source: USDA, ERS.

The world sugar price will rise 8 percent by 2009 under the PEATSIM reform scenario and 15 percent by 2010 under the U.S. Sugar and Sweetener reform scenario. The larger price effect under the latter model arises from that model's much larger decline in production and, particularly, in exports.

The significant drop in exports will reduce the supplies available on the world market. These responses are broadly comparable with those reported in other studies that use PE models, as shown in table 9. For example, Andino, Taylor, and Koo (2005) found that the combination of a 33-percent cut in the intervention price and a 15-percent cut in the production quota for EU sugar boosted the world sugar price by 14.5 percent.

Still, these magnitudes of world price increases from a unilateral EU sugar reform need to be viewed as upper-bound limits for several reasons: the price-inelastic nature of sugar markets on both the supply and demand sides; the simplifying assumption, typical of PE models, that sugar is a homogeneous good and of the same quality across suppliers and countries; and the likelihood that world price hikes will be short-term adjustments because higher prices will trigger incentives to increase sugar production, which in turn will place downward pressure on the world price. Moreover, additional dynamic effects that arise from new investments in sugar processing, along with industry rationalization and consolidation, contribute to enhanced cost efficiency and hence affect cost-price relationships.

General Equilibrium Analysis

Key differences exist between PE models and CGE models. In a CGE model, there is explicit recognition of finite resources in markets for land, labor, and capital. In doing so, there is an opportunity cost associated with a farm subsidy, where subsidizing one sector comes at the expense of other sectors because of competition for fixed resources. The PE approach introduces upward-sloping supply schedules to acknowledge higher costs as output expands. Another major difference is that in a PE model there is no link between sources and uses of income, which means there is no explicit budget constraint for the household that is tied to returns to factors. In a CGE model, it is possible to track transfers to the household for a given subsidy. Thus, the CGE approach can provide a different perspective for welfare implications of policy reform. In addition to these general differences, the GTAP Model offers specific treatment for bilateral trade between all partners. This allows for analysis of preferential and bilateral free trade reforms.

A set of scenarios was constructed to illustrate the likely impacts of different reform approaches for the EU sugar program.[9] The first scenario (table 10) implements the EBA Initiative, whereby the EU removes tariffs and quota restrictions on all imports from LDCs. The second scenario, Reform1, implements the EU reform package of November 2005, in which domestic prices are allowed to fall below an existing administered sugar price as a result of cuts in border measures for EU sugar exports and imports. The export subsidy rate is cut by 20 percent to bring subsidized exports within WTO limits. Scenario Reform1Q implements these changes plus the 15-percent production quota cut. This scenario is closest to the reform scenarios in the PEATSIM and U.S. Sugar and Sweetener Models. Finally, scenario Reform 1QE reruns Reform1Q by allowing for imperfect competition in the domestic sugar industry. EU sugar processors operate with fewer firms having market power, hence enabling a separate assessment of firm-level versus industrywide adjustments to the EU reforms. All the scenarios were modeled to generate medium-term static effects assuming exogenous total fixed supply of land, labor, and capital, but with factors mobile across sectors. Hence, any changes in sugar production came strictly from the reallocation of existing land within agriculture.

Table 9. Recent studies on EU sugar reform.

Study	Model configuration	Modeling approach	Scenarios	Main findings
Conforti and Rapso-manikis (2006)	COSIMO-AGLINK (PE); 56 regions; raw/ white sugar, beet/ cane sugar	Evaluation of reform and the EBA initiative on world markets	Baseline 2003-13; reform = 39% IP cut	5% rise in world price; lower revenue for ACP countries despite export rise; EBA exports rise threefold
El-Obeid and Beghin (2005)	CARD Sugar Model (PE); 29 regions; sugar (raw equivalent); beet/cane sugar	Removal of production and trade distortions for sugar for all countries (supply and consump tion are price inelastic)	Free trade (FT); = (1) + removal of production subsidies; = (2) + consumer distortions	FT leads to 27% higher WP; FT + no domestic distortions lead to 48% WP rise by 2012; mod-erate rise in aggregate trade
Andino, Taylor, and Koo (2005)	World sugar model (PE); 17 regions; sugar (raw equivalent)	EU sugar reform (cut in IP and PQ) and Doha Development Agenda (DDA) effect on U.S. sugar	(1)Baseline; (2)partial EU reform (IP cut; PQ cut); (3)4 country reform under DDA: (4)EU reform plus 20% tariff cut; (5)20% import quota rise for China	EU reform increases world price; no U.S. ef-fect; WP rises by 14.5%; under DDA, reform WP rises by 39%
Gohin and Bureau (2006)	Static single-region CGE model for food and agriculture for EU; 18 crops, 28 livestock commodities	In-quota and over-quota beet and sugar are distinct products; sugar is homogenenous; quota rents to exporters	(1)Baseline; (2)July 2004 EU reform + removal of export subsidies	Production falls by 10%; C-Sugar eliminated; beet prices fall (59%) more than sugar; little sugar-isoglucose substi-tution; removal of export subsidies means no imports of MFN-access and GSP but ACP coun-tries not affected
Van Berkum, Roza, and Tongeren (2005)	GTAP-derived model (CGE); 12 regions; 7 commodities	Revised land allocation; labor and capital seg-mented between agri-cultural and nonagricul-tural sectors; production quota; bilateral TRQs; sugar processing under imperfect competition; matching drop in market price with cut in export subsidies	(1)EBA Initiative; (2)EBA + July 2004 reform (beet quota cut by 16% (from 17.4 to 14.6 tons); compensation payments (transfer to sugar output; beet value added); cut in IP by 33%	Little increase in EBA imports (384,000 tons less than 2.8 million tons by European Community); mild growth of exports by ACP; modest gains for Brazil; 37% drop in exports

Notes: PE = partial equilibrium; CGE = computable general equilibrium; WP = world price; IP = intervention price; PQ = production quota; TRQ = tariff-rate quota.

Source: USDA, ERS.

Impacts of the EBA Initiative—Implementation of the EBA Initiative (scenario EBA) will result in a rise of LDC sugar exports to the EU of about $404 million (473 percent), compared with the base. However, as the initial level of this trade was small to begin with ($88 million), the effects on EU sugar production will be minor (a decrease of less than 1 percent, compared with base). Under this scenario, the impact on third countries also will be small as the trade restrictions against non-EBA partners are unchanged.

Some trade will divert from non-LDC ACP countries to the benefit of the LDCs, and countries restricted from exporting to the EU, such as Brazil, will manage to channel some additional sugar exports to the EU via the LDCs (table 11).

Since this analysis does not account for the possibility of additional production by the LDCs in response to the EBA Initiative, which was a motivating factor behind the reform initiative, some uncertainty exists as to the expected new flow of sugar from the LDCs to the EU. A sensitivity analysis helped address this uncertainty, focusing on a key variable affecting the responsiveness of export demand to policy change. This sensitivity analysis varies the Armington trade elasticity of substitution, which reflects the degree of production differentiation in sugar trade. The EBA scenario was rerun using a range of trade elasticity values from low (greater product differentiation) to high (greater homogeneity for sugar among regions).[10] The quantitative results did not change qualitatively except for the magnitude. The increase in EBA sugar exports to the EU will range from 321 percent (highly differentiated case) to 485 percent (homogenous case), compared with a 473-percent increase in the central case.[11] Translated into quantity terms and using 2001 prices, the value changes in EBA sugar exports to the EU will range from 0.50 to 0.75 million tons. These results suggest that while the LDCs may significantly boost their exports to the EU, the volume of new sugar exports may still fall short of the EU Commission's optimistic projections of 2.8 to 3.5 million tons of imports from the EBA countries by 2014 (Commission of the European Communities, 2005).

Table 10. GTAP Model scenarios.

GTAP Model scenarios	Policy change	Sugar market assumption	Starting database
EBA	Elimination of tariffs from LDCs		2005, EU-25 enlargement
EBA-S	Same as EBA	High product differentiation	
EBA-B	Same as EBA	Low product differentiation	
Reform 1	Export subsidy rate cut by 20 percent Import tariff cut by 36 percent Intervention floor price allowed to fall	Perfectly competitive industry	2005, post-EBA
Reform1Q	Reform1, plus production quota cut by 15 percent	Perfectly competitive industry	
Reform1QE	Same as Reform1Q	Imperfect competition	

Note: Export subsidy rates in GTAP are derived based on the total value of exports and the total value of subsidy outlays as reported to the WTO for the same period; for the EU, this is adjusted to take into account subsidies for ACP sugar not reported to the WTO. Product differentiation assumption is governed by trade elasticities

Source: USDA, ERS.

EU Sugar Reform: Industry Impacts

Under scenario Reform1 (domestic price cut without a production quota cut), the reform's impact on total EU sugar production will be relatively small (a decrease of 3.2 percent, table 12). In the presence of binding production quotas, sugar production will decrease only in France, Italy, and the smaller producing regions. Also, the EU market price for sugar will drop 9.2 percent in response to the reform. These output and price effects are smaller than the PE modeling results due to the general equilibrium feedback arising from the simultaneous effects of lower output (lower prices), higher imports (to meet higher demand due to lower prices), and reduced exports (through cuts in subsidies).

Table 11. GTAP Model results: EU reform and changes in bilateral trade flows.

	Base data							
		of which to:						
	Total initial exports	**EU-25**	**ACP**	**LDC**	**Mediterr-anean**	**U.S.**	**South Asia**	**Others**
	Million dollars							
EU-25	1,662.7	1,299.1	13.1	40.1	70.0	33.0	61.2	146.2
ACP	2,337.1	924.0	155.9	67.8	62.5	175.8	210.5	740.6
LDC	217.7	88.4	5.1	84.3	8.5	22.2	2.7	6.5
Brazil	1,723.9	29.2	13.7	95.3	313.5	74.6	109.9	1,087.8
C. America	944.2	31.9	41.9	9.3	9.7	239.7	119.2	492.5
India	241.5	43.4	0.3	28.6	8.9	2.7	37.2	120.5
Australia	852.2	0.7	3.1	0.0	32.2	51.1	535.6	229.5
Thailand	625.7	5.2	0.3	33.6	0.3	13.7	432.0	140.6
Others	1,537.6	435.2	23.6	25.9	63.9	337.6	272.6	378.9
World	10,142.6	1,557.0						
	Scenario EBA: Implementation of EBA for LDCs							
		of which to:						
	Total initial exports	**EU-25**	**ACP**	**LDC**	**Mediterr-anean**	**U.S.**	**South Asia**	**Others**
	Million dollars							
EU-25	3.7	-22.2	0.9	4.6	5.7	1.4	3.6	9.5
ACP	-96.2	-100.3	-0.2	2.3	0.1	0.5	0.4	0.9
LDC	402.7	404.0	-0.2	0.1	-0.3	-0.7	-0.1	-0.2
Brazil	-3.2	-3.8	-0.1	2.9	-0.3	0.0	-0.1	-1.9
C. America	-3.8	-3.3	-0.2	0.3	0.0	0.0	-0.1	-0.6
India	-2.0	-2.5	0.0	0.9	0.0	0.0	-0.1	-0.3
Australia	-1.0	-0.1	0.0	0.0	0.0	0.0	-0.6	-0.3
Thailand	0.0	-0.4	0.0	1.0	0.0	0.0	-0.5	-0.2
Others	-36.7	-36.8	-0.1	0.8	0.0	0.0	-0.2	-0.3
World	263.4	257.0						

Table 11. (Continued)

	Scenario Reform1: Support price and export subsidy cuts							
		of which to:						
	Total initial exports	EU-25	ACP	LDC	Mediterr-anean	U.S.	South Asia	Others
				Million dollars				
EU-25	-129.4	-34.4	-4.6	37.3	-24.6	-16.3	-28.3	-58.4
ACP	363.5	368.2	0.9	-1.2	-0.4	-1.3	-0.1	-2.6
LDC	-45.8	-45.6	0.1	-0.5	0.0	0.1	0.0	0.0
Brazil	36.7	31.1	0.2	-1.0	0.6	0.0	0.8	5.0
C. America	22.8	18.5	0.6	-0.1	0.0	0.2	1.0	2.5
India	8.7	8.1	0.0	-0.3	0.0	0.0	0.3	0.6
Australia	5.7	0.0	0.0	0.0	0.1	0.1	4.4	1.0
Thailand	4.8	3.0	0.0	-0.4	0.0	0.0	2.0	0.2
Others	-30.7	-34.6	0.3	-0.3	0.1	0.4	2.0	1.3
World	236.3	314.5						
	Scenario Reform1Q: Support price/export subsidy and quota cut							
		of which to:						
	Total initial exports	**EU-25**	**ACP**	**LDC**	**Mediterr-anean**	**U.S.**	**South Asia**	**Others**
				Million dollars				
EU-25	58.4	283.8	-7.9	-1.6	-45.2	-23.4	-44.2	-103.2
ACP	770.4	779.1	2.0	-1.5	-0.8	-3.0	-1.4	-4.0
LDC	9.3	8.8	0.2	0.1	0.1	0.1	0.0	0.1
Brazil	59.8	38.9	0.5	-0.2	2.2	0.2	1.5	16.7
C. America	47.2	36.6	1.4	0.0	0.1	0.7	1.7	6.7
India	65.8	64.2	0.0	-0.1	0.0	0.0	0.4	1.4
Australia	11.9	0.9	0.1	0.0	0.3	0.2	8.0	2.5
Thailand	11.6	8.6	0.0	-0.3	0.0	-0.1	2.9	0.4
Others	363.5	354.2	0.8	0.0	0.5	0.9	3.7	3.4
World	1,397.8	1,575.0						

Note: ACP = Africa, Caribbean, and Pacific countries. LDC = Least developed countries.
Source: USDA, ERS model simulation results.

Also in scenario Reform1, the decrease in the producer price (2.3 percent) will be smaller than the decrease in the market price (9.2 percent). This difference can be gleaned from the relationship linking producers to the market price via "tax" or "subsidy" wedges. Such wedges arise from (a) the sugar production quota, which, when binding, generates quota rents captured by the firm as a subsidy; and (b) the oligopolistic structure of the sugar industry, where price markups or excess profits are taxed away from the firm. In this scenario, the decrease in the producer price will be smaller than the decrease in the market price due to an increase in the tax equivalent of the production quota rent as quotas become more binding. When the reform is accompanied by a cut in production quotas (scenario Reform1Q), EU sugar output will contract by 14.3 percent (table 12), which is very close to the quota reduction of 15 percent. In this case, the market price of sugar will rise 22.5 percent, as supply is reduced. However, the producer price still contracts only by 2.3 percent due to the combined effect of the drop in quota rents (captured as a subsidy for the firm) and increased oligopoly markets (captured as a tax for the firm).

When the entry and exit of firms are permitted (scenario Reform1QE), the efficiency of the industry improves. In this scenario, the number of firms exiting the industry ranges from 13 to 17 percent across the 13 EU regions represented in the model—except for Italy, whose initial sugar output is below quota (table 13). For the remaining firms in the industry, sugar output will increase in eight regions and decrease in four, dampening the reduction in total output. The overall market price effect (not shown in tables) will be slightly larger (23.7 percent) than that of the no entry/exit case (scenario Reform 1Q). In addition, the producer price change will become slightly positive (2.9 percent) as markups over marginal cost will fall by less under free entry and exit.

Table 12. Summary of simulation results.

Item	Initial base	EBA		Reform1		Reform1Q	
		% change	**Value change[1]**	**% change**	**Value change**	**% change**	**Value change**
	Million dollars	*Million dollars*		*Million dollars*		*Million dollars*	
Imports into EU-25							
Total (Extra-EU)	**1,363.8**	**5.7**	**273.8**	**19.6**	**361.2**	**62.4**	**1,170.3**
ACP	924.0	-10.9	-100.3	45.7	368.2	92.5	779.1
LDC	88.4	473.1	404.0	-8.4	-45.6	1.6	8.8
Mediterranean	197.4	-8.7	-17.4	9.0	9.0	120.8	201.0
United States	56.4	-7.8	-4.5	-11.8	-6.8	113.6	56.4
Canada	18.7	-6.9	-1.3	-35.8	-6.4	60.8	11.0
Brazil	29.2	-12.6	-3.8	123.7	31.1	155.8	38.9
China	0.7	-8.6	-0.1	77.9	0.5	231.3	1.5
India	43.4	-5.5	-2.5	29.8	8.1	158.1	64.2
Australia	0.7	-7.9	-0.1	11.2	0.0	143.8	0.9
Thailand	5.2	-7.8	-0.4	89.7	3.0	224.8	8.6
Exports from EU-25							
Total (Extra-EU)	**183.2**	**7.1**	**14.6**	**-25.7**	**-19.6**	**-60.1**	**-97.0**
ACP	13.1	7.0	0.9	-34.4	-4.6	-58.2	-7.9
LDC	40.1	11.5	4.6	83.2	37.3	-6.4	-1.6
Mediterranean	70.0	8.2	5.7	-33.7	-24.6	-61.6	-45.2
United States	33.0	4.3	1.4	-48.1	-16.3	-70.1	-23.4
Canada	5.1	7.7	0.4	-37.5	-2.0	-70.2	-3.8
Brazil	3.3	5.3	0.2	-46.0	-1.6	-76.2	-2.6
China	3.6	8.3	0.3	-39.4	-1.5	-67.5	-2.6
India	4.5	4.9	0.2	-26.4	-1.3	-48.8	-2.2
Australia	5.2	8.3	0.4	-41.1	-2.2	-70.4	-3.8
Thailand	5.4	6.9	0.4	-46.8	-2.7	-71.5	-4.1
Output (raw sugar equivalent)							
EU-25	**26,164.0**	**-1.0**	**-251.5**	**-3.2**	**-791.0**	**-14.3**	**-3,653.3**
CP	3,981.6	-2.5	-98.8	9.6	374.0	20.4	792.6

Table 12. (Continued)

		EBA		Reform1		Reform1Q	
Item	**Initial base**	**% change**	**Value change**[1]	**% change**	**Value change**	**% change**	**Value change**
LDC	5,322.2	7.6	403.2	-1.3	-73.0	0.3	14.4
Mediterranean	16,358.7	-0.1	-23.1	0.2	32.5	1.7	269.2
United States	32,477.9	0.0	-8.6	0.0	12.5	0.4	121.5
Canada	331.4	-0.5	-1.5	-1.9	-6.3	4.2	13.8
Brazil	5,800.6	-0.1	-4.1	0.8	45.1	1.3	74.4
China	412.7	-0.1	-0.3	0.4	1.7	1.0	4.2
India	7,171.9	0.0	-2.5	0.1	9.7	1.0	68.1
Australia	2,196.4	-0.1	-1.9	0.5	9.8	0.9	20.2
Thailand	1,527.6	0.0	-0.3	0.5	7.1	1.0	15.7
Income (millions of 2001 USD)							
EU-25	**8,988,151.1**		**-317.2**		**1,302.0**		**1,600.7**
ACP	385,024.0		-21.0		-19.1		79.9
LDC	344,329.3		362.9		-14.5		-8.3
Mediterranean	1,178,657.0		3.5		4.1		24.3
United States	11,057,000.0		-39.6		-13.7		48.4
Canada	780,615.4		4.2		-2.3		7.4
Brazil	595,336.6		1.8		2.4		27.0
China	1,384,773.0		-6.2		-2.3		-4.1
India	567,009.9		-12.6		-1.7		10.4
Australia	400,031.3		5.8		-2.4		9.8
Thailand	142,727.5		0.4		-2.4		5.5
Rest of world	1,310,200.1		7.7		-3.5		-11.4
Sugar quota rent transfer by EU-25	1,151.4		135.1		-478.6		-455.7

Note: ACP = Africa, Caribbean, and Pacific countries. LDC = Least developed countries. [1]Value change is expressed in constant 2001 U.S. dollars.
Source: USDA, ERS model simulation results.

On the demand side, the price changes resulting from the reform will lead to a small increase (1.1 percent) in EU sugar consumption under Reform 1 and no change in consumption under Reform1Q, as the EU's demand for sugar is very inelastic. However, in the CGE analysis, the substitution between sugar and isoglucose is not directly modeled because the possibilities for substituting sweeteners for sugar remain limited in the EU. Unlike in the United States, where the soft drink industry relies heavily on HFCS, the EU soft drink industry uses mainly saccharose and very little isoglucose, hence, limiting potential substitutions with sugar. A separate general equilibrium study of the EU sugar reform that used a model to estimate sugar-isoglucose substitution (Gohin and Bureau, 2006) found little increase in consumption from cuts in sugar prices.

These scenarios suggest that significant cuts in EU sugar output may not be realized solely through price support cuts, given the oligopolistic nature of the EU sugar industry. In 2006 and 2007, a significant restructuring of the industry took place, with both plant closures and quota reallocation between plants and countries. At least 7 factories closed in Germany; the number of plants in Italy is expected to decline from 19 to 6; all of Ireland's factories have closed; and production in Austria, Denmark, and Belgium is now concentrated in only 2 plants each (Bureau et al., 2007). However, the net decline in quota production resulting from this rationalization was just over 1 million tons, or 5.4 percent of the total production quota in 2005. Thus, a mandatory cut in production quotas may be unavoidable if the EU is to stay within its lowered sugar production targets. This is, in fact, what happened, as the EU Commission reduced production quotas in 2006 and 2007. In May 2007, the Commission decided to withdraw about 2 million tons, or 10.8 percent of total sugar quota allocations for the EU, leaving only 16.6 million tons for 2007/08.

Trade and Income Impacts

The model results indicate that cuts in the domestic price of sugar and sugar production quotas will lead to much larger adjustments in production and trade than a price cut alone. Under scenario Reform 1 (price cut only), EU sugar exports will drop by 25.8 percent, the change in EU sugar production will decrease by 2.3 percent, and EU sugar imports will increase by 19.6 percent (see table 12). In contrast, under scenario Reform1Q (price and quota cut), the changes in exports, production, and imports are all far more substantial: exports will decrease by 60.1 percent, production contracts will decrease by 14.5 percent, and imports will increase by 62.3 percent. Consequently, the volume of world sugar trade increases by only 2.5 percent under scenario Reform1, compared with 14.5 percent under scenario Reform1Q.

The impact of the EU reform on world prices will be small, ranging from a slight drop (0.85 percent) under scenario Reform1 to an increase of about 1 percent under scenario Reform1Q. The small decrease in the world price under Reform 1 is due to the lower price of EU sugar imports and lower export prices as the EU cuts export subsidies. When EU price changes are weighed against the modest price changes in other countries, the net result will be a small overall change in the world price index. Under scenario Reform1Q, the change in the world price will be slightly positive, as the large gains in EU sugar imports boost the benefit of exporting. The net result will be a slight rise in the world price due to the reduction of EU sugar exports.

The rise in EU sugar imports will benefit primarily the ACP countries, which will expand their exports beyond the quotas of the former Sugar Protocol. Because the reform scenario already assumes that EBA-eligible LDCs have duty- and quota-free access to the EU market, implementation of the reform will add no new benefit to these countries. On the contrary, the EBA-eligible LDCs will face greater competition from non-LDC suppliers among the ACP countries. For the non-EBA and non-ACP sugar suppliers to the world market, exports will grow only modestly as world price changes will not be large enough to affect the world demand for sugar. Brazil and Central America will benefit some, but Australia and Thailand, which export sugar mostly to Asian countries, will see minimal changes.

Welfare and income effects for the LDCs under the EBA Initiative (without the EU sugar reform) show that income gains will come largely from tariff removal on nonsugar exports because initial sugar exports to the EU are relatively small, compared with those of other commodities. The EU will see a small welfare loss under the EBA Initiative, due mostly to lost tariff revenues. Under the reform scenarios, however, the EU's welfare change is positive due to the interplay of several factors: lower sugar prices and production costs for sugar-using industries; benefits to consumers through cheaper imports; contraction of the sugar industry leading to smaller deadweight losses; and lower export subsidies.

Table 13. EU reforms and market structure for EU sugar industry (GTAP Model results).

	Change in number of firms	Change in output per firm	Change in markups	Change in industry output	Change in industry output	Change in exports	Change in imports	Change in regional income
	Percent				*Million dollars*			
	Reform1 Q scenario (no firm entry or exit)							
Belgium	0	-15.00	4.51	-15.00	-227	-27	159	83
Denmark	0	-15.00	5.00	-15.00	-58	-4	106	5
Germany	0	-15.00	1.39	-15.00	-771	-32	1,534	-55
Spain	0	-15.00	4.83	-15.00	-145	-17	223	32
France	0	-15.00	-0.82	-15.00	-488	-48	863	305
Italy	0	-2.10	0.02	-2.10	-28	-31	67	53
Netherlands	0	-15.00	1.91	-15.00	-109	-21	114	37
Austria	0	-15.00	5.00	-15.00	-67	-3	135	13
Poland	0	-15.00	4.60	-15.00	-863	-7	7,398	-183
Sweden	0	-15.00	4.59	-15.00	-88	-6	139	-5
United Kingdom	0	-15.00	5.03	-15.00	-509	-13	366	885
Hungary	0	-15.00	4.77	-15.00	-45	-1	68	1
Czech Republic	0	-15.00	4.54	-15.00	-133	-1	205	-29
Other EU	0	-14.65	5.34	-14.65	-320	-18	460	459
	Reform 1QE scenario (with free firm entry or exit)							
Belgium	-15.42	0.50	0	-15.00	-229	-27	199	20
Denmark	-15.15	0.18	0	-15.00	-58	-4	108	-10
Germany	-13.64	-1.57	0	-15.00	-772	-32	1,550	-118
Spain	-15.05	0.06	0	-15.00	-145	-17	224	-8
France	-17.33	2.81	0	-15.00	-489	-48	869	325
Italy	-0.11	-1.86	0	-1.96	-26	-31	67	51
Netherlands	-16.13	1.35	0	-15.00	-109	-21	116	26
Austria	-15.15	0.18	0	-15.00	-67	-3	136	-7
Poland	-16.11	1.32	0	-15.00	-890	-6	10,064	-289
Sweden	-14.88	-0.14	0	-15.00	-89	-6	145	-25
United Kingdom	-14.96	-0.04	0	-15.00	-512	-11	413	727
Hungary	-15.08	0.09	0	-15.00	-45	-1	72	-10
Czech Republic	-14.71	-0.34	0	-15.00	-134	-1	216	-55
Other EU	-14.18	-0.34	0	-14.48	-317	-18	472	344

Source: USDA, ERS model simulation results.

For preferential partners, the LDCs will show no additional benefit from the reforms (in the post-EBA environment), while the change in welfare in non-LDC ACP countries will depend on the extent of the EU sugar industry's increase in efficiency. Under modest contraction of sugar output in the EU (scenario Reform1), the welfare change for the ACP countries will be negative despite increased exports. This effect arises due to lower border-cumdomestic prices in the EU. These prices translate to quota rent losses, which dominate the welfare change. By contrast, under sufficient EU sugar production cuts, the trade creation for the ACP countries arising from larger sugar exports will outweigh the quota rent losses, resulting in a net welfare gain for the ACP countries. For nonpreferential exporters that manage to increase their exports under scenario Reform1Q, income will rise but only slightly.

Finally, given the complexity of the EU sugar regime, it is important to note some caveats inherent in the model assumptions. First, given the long-standing interventionist nature of EU sugar policies, it is not easy to calibrate the proper supply response for EU sugar. This difficulty arises in all quantitative models that deal with EU sugar (Bureau et al., 2007). Second, the model does not include dynamics such as capital accumulation that would mimic future investments in sugar processing in third countries benefiting from the EBA Initiative. Such investments are likely to push EU sugar imports beyond the levels reported in this study's comparative static analysis. Finally, the model does not explicitly address several elements of the reform, such as farm payments and payments for quota buyout. Inclusion of these elements would have rendered the analysis excessively complicated and difficult to interpret. Despite these caveats, many critical features of the previous EU sugar regime and the reform were incorporated, and the model results seem both intuitive and defensible.

CONCLUSIONS

The 2005 EU sugar reform, the first in nearly 40 years, has important implications for EU and international sugar markets. The reform is an effort to mitigate the expected increase in domestic imbalances in supply and demand arising from larger sugar imports and greater constraints on exports. Several developments are interpreted as signs of this expected imbalance: (1) the likely rise in sugar imports from least developed countries following the implementation of the EBA Initiative, (2) the issuance of new access quotas to compensate third-country exporters for the EU expansion in 2004, and (3) a WTO panel decision declaring some EU sugar export subsidies to be incompatible with the EU's WTO commitments.

The EU's CMO for sugar is a complex mix of policy instruments that includes price support, production quotas for sugar and sugar substitutes, export subsidies, and import barriers. However, the reform targeted only a few of these instruments, principally cutting the intervention price 36 percent to discourage domestic production and sugar imports from the LDCs under the EBA Initiative. In addition, the reform included a voluntary buyout scheme for production quotas and the disallowance of exports of nonquota sugar to rein in domestic production and to bring export subsidies within the WTO limits.

Other policies that would have enhanced internal competition and made the EU sugar industry more efficient were not included in the reform. For example, interstate quota trading, which could induce a significant shift in production from high- to low-cost regions, was not included, leaving in place the allocation of production quotas on a national basis. Also, restrictive production quotas on sugar substitutes remain in place, albeit at higher levels, preventing greater competition among different types of sweeteners. Despite substantial concentration in the EU sugar industry and the industry's noncompetitive pricing, the reform (which does not lower import barriers) does nothing to affect either industrial concentration or price transparency. Consequently, market prices of sugar, which tend to operate at levels above intervention prices, may fall by less than the cut in the intervention price.

The analysis suggests that, given the market structure of the EU sugar industry, cutting the intervention price alone may not have the desired effect on market prices and production. The reform's impact on production likely will depend more on the amount of sugar quota that is removed from the market. This, in turn, will depend on the extent of industry consolidation and restructuring. The first 2 years of the reform (2006 and 2007) have been marked by a wave of factory closures and reallocation of production across plants and countries. This restructuring generally did not result in individual member states abandoning a sufficiently large quota volume to reduce production to levels anticipated by the EU Commission. Rather, the member states have reallocated their quota volumes internally among their own processing plants. Thus, the net decline in quota production was smaller than expected— just over 1 million tons, or 5.4 percent of total quota production in 2005. Subsequently, the EU Commission made quota renouncement more attractive and bought out more than 1.5 million metric tons, although the Commission expects that a permanent cut of over 1 million metric tons will be necessary in 2010 to balance EU supply with demand.

For sugar beet producers, the cuts in the intervention prices, combined with a more lucrative buyback of quota, may induce lower sugar output if these measures make other economic opportunities more attractive.

Producers still can shift some sugar beets to ethanol production and collect the energy subsidy. Moreover, given the rising cereal prices induced by expanded demand, sugar beet producers could shift to cereals—mostly wheat (typically grown in rotation with sugar beets), particularly in regions with higher production costs. Nevertheless, a more complete sugar reform that includes lower import barriers (outside preferential agreements), inter-country quota tradability, and greater market opportunities for sugar substitutes would improve the efficiency of the industry by weakening oligopolistic behavior, enhancing internal competition, and providing a larger scope for output contraction.

Trade effects of the EU sugar reform suggest that the concomitant implementation of the EBA Initiative in 2009 and the EU reform (which started in 2006) will dampen the gains for LDCs as they face greater competition from other exporters to the EU. For non-LDC exporters among the African-Caribbean-Pacific (ACP) countries, the possibility of increased sugar exports to the EU may be tempered by lost income due to preference erosion and the lower quota rents associated with reduced support prices received by preferential exporters. However, for nonpreferential exporters, the direct implica-tions of the EU reform for bilateral trade are small, owing to the dominance of the EU's preferential sugar agreements.

The EU sugar reform is expected to exert upward pressure on world sugar prices, but such price increases are likely to be overtaken by much larger price hikes due to rising demand for sugar in order to produce ethanol, mostly from Brazil. Increasing use of sugarcane for Brazilian ethanol production means tighter world sugar supplies and higher prices. This development could help many of the ACP sugar producers who face an erosion of preference benefits from the EU sugar policy reforms. Sugar exporters from the ACP countries and the LDCs would not be the only beneficiaries of rising world sugar prices. Some countries could become producers of sugarcane- based ethanol and contribute to the expanding global biofuel market.

Finally, the reform could have important implications for the WTO. While the reform reduces EU sugar export subsidies to within the current WTO limits, it also could serve as the first step in a phased removal of these subsidies under the Doha Development Round. In the area of domestic support, cuts in the intervention price and lower sugar production may reduce the EU's aggregate measure of support. However, in the area of market access, the reform does not address the high import barriers outside preferential agreements. The ad valorem level of these barriers is estimated at more than 160 percent. Significant tariff cuts for sugar of about 60-90 percent, like those tabled in the 2005 WTO meeting in Hong Kong, would not only have a greater impact on world sugar markets but would affect EU sugar trade with nonpreferential trading partners.

References

[1] Andino, J., Taylor, R. & Koo, W. (2005). "Impact of the EU Sugar Policy Reform and the WTO-DDA on the U.S. Sugar Industry," paper presented at the Annual *Meeting of the American Agricultural Economics Association,* Providence, RI, July 24-27.

[2] Agra Europe (2004). *CAP Monitor – Update Issue No. 11: Sugar,* London. Agra Europe (2007). *EU Agrees To Scrap ACP Sugar Protocol*, October 5. Agra Europe (2008). *EU Sugar Producers Renounce More Quota*, January 22.

[3] Berkum, S., Roza, P. & Tongeren, F. (2005). *Impacts of the EU Sugar Policy Reforms on Developing Countries*, Report No. 6/05/09, Agricultural Economics Research Institute (LEI), The Hague.

[4] Bruhns, J., Baron, O. & Maier, K. (2004). "Sugar and Sweetener Economy: Europe and North America 2005," *Bartens,* Berlin.

[5] Bureau, J. C., Gohin, A., Guide, L. & Millet, G. (2007). "*EU Sugar Reforms and Their Impacts*," paper presented at the International Food Policy Research Institute conference entitled "Global Sugar Markets, Policies and Reform Options," June 1, Washington, DC.

[6] CAP Monitor (2004). *The Complete Guide to the Common Agricultural Policy*; Agra Europe (London) Ltd.

[7] Chaplin, H. & Matthews, A. (2006). "Coping With the Fallout for Preference-Receiving Countries From EU Sugar Reform," *The Estey Centre Journal of International Trade and Policy* 7(1):15-31, available at: http://www.esteycentre.com/journal/j_ pdfs/chaplinmatthews7-1.pdf

[8] Commission of the European Communities (2005). *Reforming the European Union's Sugar Policy: Update of Impact Assessment*, Brussels, June.

[9] COMTRADE (2004). *United Nations Commodity Trade Statistics Database*, United Nations Statistics Division.

[10] Conforti, P. & Rapsomanikis, G. (2006). "Preferences Erosion and Trade Costs in the Sugar Market: The Impact of the Everything but Arms Initiative and the Reform of the EU Policy," paper presented at the 26th Conference of the International Association of Agricultural Economics, Gold Coast, Australia, August.

[11] Dimaranan, B. & MacDougall, R. (eds.)(2006). *Global Trade, Assistance, and Production: The GTAP 6 Data Base.* Center for Global Trade Analysis, Purdue University.

[12] Elbehri, A. & Hertel, T. (2006). "A Comparative Analysis of the EU-Morocco FTA vs. Multilateral Liberalization," *Journal of Economic Integration, 21*(3), 496-525.

[13] Elobeid, A. & Beghin, J. (2005). "Multilateral Trade and Agricultural Policy Reforms in Sugar Markets," Working Paper No. 04-WP 356, Center for Agricultural and Rural Development, Iowa State University.

[14] Francois, J. (1998). *Scale Economies and Imperfect Competition in GTAP Model*, Global Trade Analysis Project (GTAP) Technical Paper No. 14, Purdue University.

[15] Frandsen, S., Jensen, H., Wusheng, W. & Walter-Jorgensen, A. (2003). "Reform of EU Sugar Policy: Price Cuts Versus Quota Reductions," *European Review of Agricultural Economics, 30*, 1-26.

[16] Gohin, A. & Bureau, J. C. (2006). "Modeling the EU Sugar Supply to Assess Sectoral Policy Reforms," *European Review of Agricultural Economics, 33*(2), 223-247.

[17] Haley, S. (2006). "European Union-25 Sugar Policy," In: S., Haley, D. Kelch, & A. Jerardo, *Sugar and Sweeteners Outlook*, Outlook Report No. SSS-245, U.S. Department of Agriculture, Economic Research Service, January 21:23-36, available at: http://usda.mannlib.cornell.edu/usda/ers/ SSS//2000s/2006/SSS-0 1-3 1-2006.pdf.

[18] Harrison, G., Rutherford, T. & Tarr, D. (1994). "*Product Standards, Imperfect Competition and the Completion of the Market in the European Community*," Policy Research Working Paper No. 1293, World Bank.

[19] Hertel, T. (ed.)(1997). *Global Trade Analysis*, Cambridge University Press.

[20] Hertel, T., Hummels, D., Ivanic, M. & Keeney, R. (2003). "*How Confident Can We Be in CGE-based Assessments of Free Trade Agreements?*" Purdue University, Global Trade Analysis Project (GTAP) Working Paper No. 26, available at: www.gtap.org.

[21] Huan-Niemi, E. & Niemi, J. (2003). "*The Impact of Preferential, Regional and Multilateral Trade Agreements: A Case Study of the EU Sugar Regime*," ENAPRI Working Paper No. 1.

[22] Larson, D. & Borrell, B. (2001). "*Sugar Policy Reform*," Policy Research Working Paper No. 2602, World Bank.

[23] Netherlands Economic Institute (2000). *Evaluation of the Common Organization of the Markets in the Sugar Sector*, Agricultural Economics and Rural Development Division.

[24] Organisation for Economic Co-operation and Development (2007). *Sugar Policy Reform in the European Union and World Sugar Markets*.

[25] Roland-Holst, D., Reinert, K. & Shiells, C. (1994). "A General Equilibrium Analysis of North American Economic Integration," in Francois, *J. and C. Shiells,* (eds.), *Modeling Trade Policy: AGE Assessments of North American Free Trade*, Cambridge University Press.

[26] Stout, J. & Abler, D. (2003). "ERS/Penn State Trade Model Documentation," U.S. Department of Agriculture, Economic Research Service, and Penn State University, Department of Agricultural Economics and Rural Sociology, available at: http://trade.aers.psu.edu/ index.cfm.

[27] Swedish Competition Authority (2002). *Sweet Fifteen: The Competition on the EU Sugar Markets*, Swedish Competition Authority Report No. 7.

[28] Swinbank, A. (2004). "Dirty Tariffication Revisited: The EU and Sugar," *The Estey Centre Journal of International Law and Trade Policy*, *Vol. 5*, No. 1, 56-69.

[29] World Trade Organization (2005). *European Communities – Export Subsidies on Sugar: Report of the Appellate Body*, AB-2005-2, April 28.

End Notes

[1] The one major exception occurred in 1995, when the EU agreed to reduce its export subsidies by 21 percent in volume and 36 percent in value over the period 1995-2001 as part of the WTO Agreement on Agriculture (AoA).

[2] The accession of Romania and Bulgaria, effective January 2007, is not included in the present analysis.

[3] As part of the WTO AoA, members were required to maintain current import access opportunities for tariffied products at levels corresponding to those existing during the 1986-88 base period. Where such "current" access had been less than 5 percent of domestic consumption of the product in ques-tion in the base period, an (additional) minimum access opportunity had to be opened on a Most-Favored-Nation basis. This requirement was to ensure that in 1995, current and minimum access opportunities combined represented at least 3 percent of base-period consumption.

[4] All measurements in tons in this chapter refer to metric tons.

[5] The 1.8 million tons are expressed as white sugar and are split among the UK (1.0 million tons), France (0.30 million tons), Portugal (0.30 million tons), Finland (0.060 million tons), and Slovenia (0.02 million tons).

[6] According to Article 5.5 of the WTO Agreement on Agriculture, given the trigger price P, if the import price in CIF terms is X and $0.9P > X > 0.60P$, then the additional safeguard duty is $(0.9P - X) * 0.3$. Other formulas apply to other price ranges.

[7] Results from the U.S. Sugar and Sweetener Model are taken from Haley (2006).

[8] The model specifies asymmetric production responses to sugar price changes. This means that processing production capacity reductions are irreversible—closed factories do not re-open.

[9] An updating procedure was applied to the initial GTAP database (version 6.5) to change the base year from 2001 to 2004 using actual GDP and population growth rates for all countries in the world.

[10] This analysis uses the value of 5.4, following recent econometric work by Hertel et al. (2003) on trade elasticities.

[11] Van Berkum, Roza, and Tongeren (2005) also applied a CGE model to ex-amine the impact of the EBA Initiative. They found that LDC exports to the EU increase by 384,000 tons, but that quantity rises to 900,000 tons under a more homogeneous assumption.

In: Sugar: Background, Policies and Issues
Editor: Cecilia M. Jordan
ISBN: 978-1-61761-116-2

Chapter 5

SUGAR AND SWEETENERS OUTLOOK (U.S. SUGAR MAY 2010)*

Stephen Haley*† *and Mike McConnell

U.S. SUGAR MAY 2010

U.S. sugar supply for fiscal year (FY) 2011 is projected to be 3.3 percent less than in FY 2010. Although production in FY 2011 is expected to be higher than in FY 2010 by 318,000 short tons, raw value (STRV), lower beginning stocks (269,000 STRV less) and lower imports (435,000 STRV less) are more than offsetting. Higher beet sugar production reflects a return to trend yields, while cane sugar production is increased for Florida and Texas. Imports under the tariff-rate quota (TRQ) are projected at the minimum of U.S. commitments to import raw and refined sugar, including an expected shortfall. The Secretary of Agriculture will establish the actual level of the TRQ at a later date. Imports from Mexico are nearly unchanged. Total use (combined deliveries, exports, and refining losses) is unchanged.

Mexico's FY 2011 sugar supply is up 1.7 percent, with higher stocks and production more than offsetting lower imports. Production is projected to increase, as yields rebound to trend levels. Imports reflect mainly U.S. exports for Mexico's IMMEX program. Domestic sugar consumption is down slightly, due to higher expected use of corn-based sweeteners, and exports are up slightly. Ending stocks are moderately above a year earlier.

For FY 2010 U.S. sugar, increased supplies nearly offset increased use, compared with a month earlier. Imports are increased under the TRQ, in line with USDA's April 23, 2010, announcement and expectations for increased imports at the high-tier tariff. Cane sugar production is reduced. Sugar deliveries are increased to reflect the recent strong pace to date and improved data availability that made possible better analysis of recent trends.

* This is an edited, reformatted and augmented edition of a United States Department of Agriculture publication, Report SSS-M-261, dated May 13, 2010.

† Corresponding author: shaley@ers.usda.gov, mmcconnell@ers.usda.gov

U.S. SUGAR

On May 11, 2010, the U.S. Department of Agriculture (USDA) released its latest supply and use estimates for fiscal year (FY) 2010 and first projections for FY 2011 in the *World Agricultural Supply and Demand Estimates* (WASDE) report.

Production

For most of the year, the USDA's Interagency Commodity Estimates Committee (ICEC) for sugar does not project sugar production for the out-year crop. For the most part, the USDA accepts the production estimates and projections provided by beet sugar processors and cane sugar millers to the Farm Service Agency. However, the processors' and millers' forecasts are not available until June of the preceding crop year. Therefore, in the meantime, WASDE reflects ICEC projections for FY 2011 sugar in May and June.

FY 2011 Beet Sugar Production

The National Agricultural Statistics Service (NASS) forecasts sugar beet acreage intentions for crop year 2010/11 at 1.174 million acres, about 9,000 acres, or 0.8 percent, lower than area planted in crop year 2009/10. Areas planted in all regions are close to last year's levels, with Michigan and the Far West (Idaho, Oregon, and California) slightly above and the Upper Midwest (Minnesota and North Dakota) and Great Plains (Colorado, Montana, Nebraska, and Wyoming) slightly below.

In making its sugar beet production forecast, the sugar ICEC assumes normal sucrose levels and normal area harvested-to-planted ratios. The national sugar beet yield is forecast at 26.68 tons/acre, just about equal to the 2008/09 record yield of 26.71 tons/acre. Area planted to genetically modified organism (GMO) seed varieties is forecast at 95 percent of total area, about the same as last year. The jump in yields in most States took place in 2006/07 and was due mainly to use of rhizomania-resistant seed varieties and use of Pancho Beta to control for Curly Top. Fuller adaption of GMO seeds reinforces the trend of substantially higher yields into the coming year (figure 1).

Assuming trend improvements in processing, the USDA projects national beet sugar production in crop year 2010/11 at 4.630 million short tons, raw value (STRV). Figure 2 shows the direct relationship between sugar beet yield and sugar production used for prediction. The modeled relationship holds for the 1981/82-2009/10 period but recognizes that weather-related disturbances can affect the relationship, as that which occurred in 2009/10 when difficult harvest conditions affected sucrose recovery.[1]

FY 2011 Cane Sugar Production

The USDA projects FY 2011 cane sugar production at 3.535 million STRV, an increase of 187,000 STRV over that of FY 2010. Because area harvested is not forecast by NASS until the end of June, current cane sugar projections assume the same area harvested for sugar as in the previous year. Florida cane sugar production for FY 2010 is forecast at 1.785 million STRV, assuming normal weather. Unfortunately, weather conditions in Florida have been off the norm in recent years: possible freeze damage in FY 2010, remarkably dry conditions for several years, and hurricane damage to the crop in FY 2006. The USDA projects FY 2011 Louisiana cane sugar production at 1.465 million STRV, the same as this fiscal year. The USDA projects Texas production at 150,000 STRV, implying recovery from this year's poor harvest conditions. Hawaiian production is forecast at 135,000 STRV. There is now only one sugar processor in Hawaii on the island of Maui.

Trade

Although the raw and refined sugar tariff-rate quotas (TRQ) for FY 2011 have not yet been announced, the USDA projects them at minimum levels implied by existing international commitments to the World Trade Organization (WTO) and at the allocated levels of existing Free Trade Agreements (FTAs). The projection in the May WASDE is, therefore, at 1.224 million STRV (table 1). Projected shortfall is forecast at 160,257 STRV.

Until the TRQ is announced, there is no projection for additional specialty sugar. This sugar is mostly organic sugar, and its allocation for FY 2010 was set at 75,000 STRV in addition to the 1,825 STRV included in the minimum access quantity. The USDA projects imports from Mexico at 550,000 STRV, a level close to this year's estimate of 540,000 STRV. Projected exportable sugar supply in Mexico is based on an assumed return to normal production levels in FY 2011 and on expanded use of high fructose corn syrup (HFCS) in Mexico (see below).

Other program sugar imports outside the sugar TRQ for FY 2011 are projected to total 300,000 STRV. Other USDA import programs include the Refined Sugar Re-export Program, the Sugar-Containing Products Program, and the Polyhydric Alcohol Program. High-tier tariff sugar imports and sugar in imported syrups are projected at 10,000 STRV. Sugar exports for FY 2011 are forecast at 150,000 STRV. Most of these exports are expected to go to Mexico, where they are used in Mexico's product re-export (IMMEX) program. Almost all such sugar-containing products are expected to be exported to the United States.

On April 23, 2010, the Secretary of Agriculture announced an additional in-quota quantity for the raw sugar TRQ for FY 2010 in the amount of 200,000 STRV (181,437 metric tons, raw value (MTRV)). The country-by-country allocations were announced by the Office of the U.S. Trade Representative (USTR) on May 5. The total sugar TRQ for FY 2010 is estimated at 1.554 million STRV, an increase of 167,000 STRV over the April 2010 projection. TRQ shortfall is projected at 100,256 STRV. Also, the estimate for high-tier tariff imports for FY 2010 was increased from 10,000 STRV to 75,000 STRV. These imports of refined sugar at the high-tier tariff began in April and are expected to continue through September 2010.

Deliveries, Refining Losses, and Ending Stocks

Deliveries for domestic food and beverage use for FY 2011 are projected at 10.500 million STRV, the same level as for FY 2010. The FY 2010 estimate was raised by 160,000 STRV based on strong deliveries in March 2010 and also on an improved set of data from the

Foreign Agricultural Service on sugar imports from Mexico. Refining losses are expected to be in the area of -200,000 STRV for both FY 2010 and FY 2011.

Ending stocks are the difference between total supply and total use. For FY 2010, ending stocks are estimated at 1.230 million STRV, implying an ending-year stocks-to-use ratio of 11.6 percent. The ending-year FY 2010 stocks estimate is the beginning stocks projection for FY 2011. Ending stocks for FY 2011 are projected at 844,000 STRV. The implied ending-year stocks-to-use ratio is 7.9 percent.

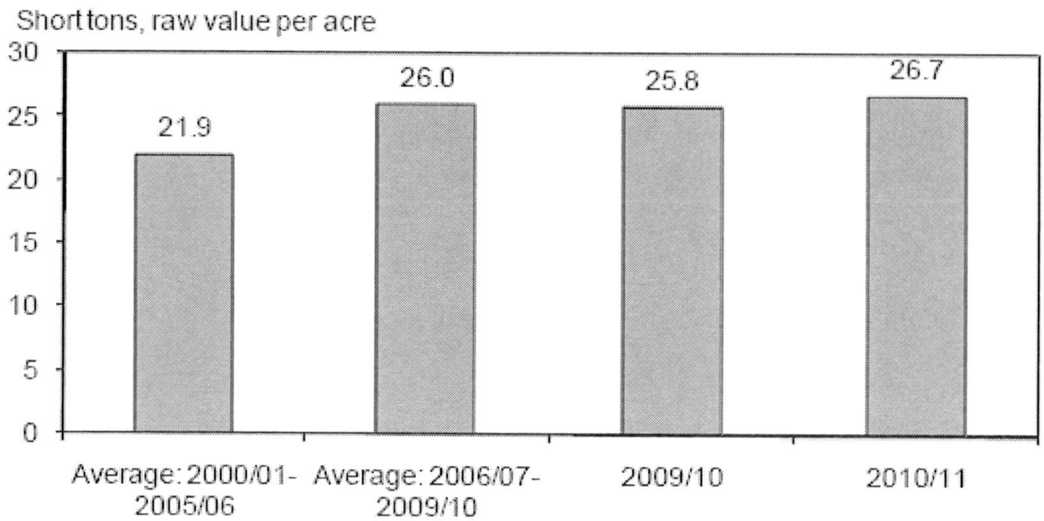

Source: Actual -- USDA, National Agricultural Statistics Service, *Crop Production*; Projection: USDA, ERS, Sugar and Sweeteners Team.

Figure 1. Sugarbeet yield, national, actual (2000/01-2009/10) and projected 2010/11.

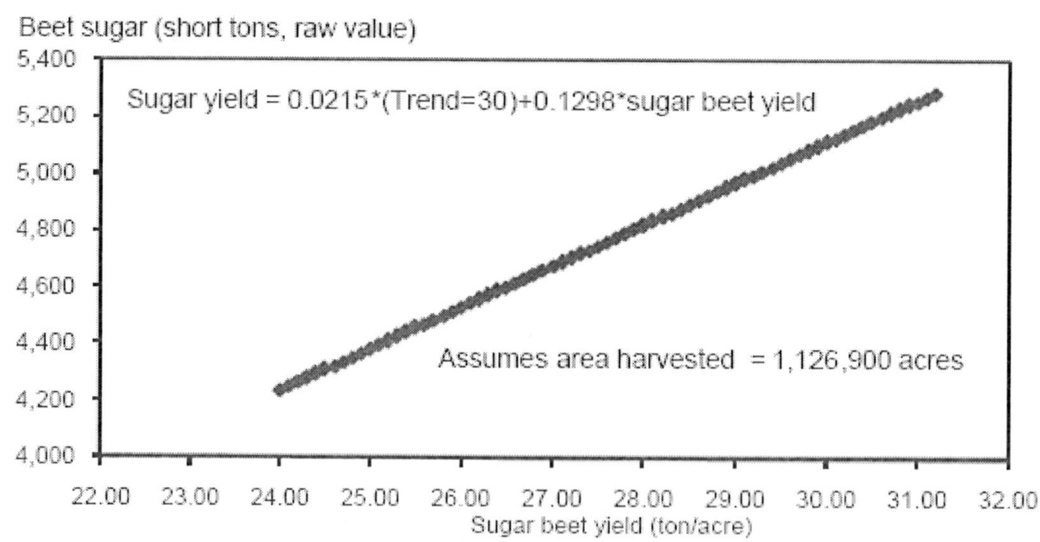

Source: USDA, ERS, Sugar and Sweeteners Team.

Figure 2. Relationship between sugar beet yield and projected beet sugar production for 2010/ 11.

Table 1. USDA estimate of sugar imports in FY 2011 5/13/10.

Item	Metric tons, raw value	Short tons, raw value
Raw sugar TRQ	1,117,195	1,231,497
Less shortfall attributable to Mexico 1/	-7,258	-8,001
Less other shortfall	-135,171	-149,000
Total raw sugar TRQ	974,766	1,074,496
Refined sugar TRQ		
Allocation to Canada	10,300	11,354
Allocation to Mexico	2,954	3,256
Less Mexican shortfall 1/	-2,954	-3,256
Global	7,090	7,815
Specialty		
Base	1,656	1,825
Additional	0	0
Total refined sugar TRQ	19,046	20,994
CAFTA/DR TRQ - calendar 2010	114,700	126,434
Singapore, Bahrain, Jordan	54	59
Peru	2,000	2,205
Total estimate TRQ entries	1,110,566	1,224,188
Mexico	498,957	550,000
Re-export program imports	272,158	300,000
Sugar syrups, high-tier	9,072	10,000
Total projected imports	1,890,753	2,084,188

TRQ = Tariff-rate quota.
CAFTA/DR = Central American Free Trade Agreement/Dominican Republic.
1/ Total entries from Mexico, quota and nonquota, reflected below.
Source: USDA, Foreign Agricultural Service.

U.S. Sweetener Demand

The Economic Research Service (ERS) makes calendar-year estimates of total sweetener deliveries that are available for food and beverage consumption by U.S. consumers. These sweeteners include refined sugar; the corn sweeteners of HFCS, glucose syrup, and dextrose; honey; and other edible syrups, including maple syrup and maple sugar. Table 2 shows the new estimates for 2009 and also includes refined sugar revisions for 2007 and 2008.

U.S. deliveries of total sweeteners for human food and beverage use for 2009 are estimated at 20.078 million tons, representing a decrease of 3 percent compared with deliveries in 2008. Refined sugar deliveries decreased by 1.8 percent, and corn sweetener deliveries fell by 4.1 percent. Within the corn sweetener category, HFCS deliveries fell by 4.7 percent. Since 2002, HFCS deliveries have fallen by 1.35 million tons, dry weight, or 14.9 percent. Honey deliveries decreased by 6.7 percent, and other edible syrups increased slightly.

Table 2. U.S. caloric sweeteners estimated deliveries for domestic food and beverage use, by calendar year 1/ 5/13/10.

Calendar year	U.S. population 2/ (July 1) Millions	Refined		Corn sweeteners			Pure honey	Edible syrups	Total caloric sweeteners	Sugar from net imports 3/	Total caloric sweeteners with sugar-containing products
		sugar	HFCS	Glucose syrup	Dextrose	Total					
1,000 tons, dry basis											
1992	256.9	8,249	6,652	1,943	461	9,056	126	53	17,483	174	17,657
1993	260.3	8,305	7,086	2,050	481	9,617	135	56	18,112	183	18,295
1994	263.4	8,478	7,398	2,093	502	9,993	126	54	18,651	153	18,804
1995	266.6	8,652	7,676	2,176	528	10,380	120	57	19,209	121	19,331
1996	269.7	8,785	7,788	2,216	537	10,541	131	57	19,514	116	19,630
1997	272.9	8,861	8,240	2,364	511	11,116	129	58	20,163	197	20,360
1998	276.1	8,966	8,552	2,358	502	11,411	130	59	20,566	274	20,840
1999	279.3	9,264	8,897	2,281	488	11,666	147	60	21,138	354	21,492
2000	282.3	9,253	8,845	2,230	476	11,551	157	61	21,022	304	21,327
2001	285.0	9,195	8,920	2,205	469	11,595	134	61	20,986	388	21,373
2002	287.7	9,109	9,045	2,224	473	11,741	153	62	21,065	529	21,594
2003	290.3	8,859	8,849	2,209	449	11,507	146	63	20,575	621	21,196
2004	293.0	9,045	8,779	2,292	487	11,558	130	64	20,797	656	21,453
2005	295.7	9,346	8,756	2,261	481	11,497	155	66	21,065	669	21,734
2006	298.4	9,347	8,702	2,053	463	11,218	174	66	20,806	856	21,662
2007	301.1	9,381	8,479	2,067	448	10,994	141	67	20,583	836	21,419
2008	303.8	9,937	8,080	2,036	419	10,535	150	68	20,690	649	21,340
2009 4/	307.2	9,763	7,698	1,991	417	10,105	140	70	20,078	555	20,633
Pounds, dry basis											
1992	----	64.2	51.8	15.1	3.6	70.5	1.0	0.4	136.1	1.4	137.5
1993	----	63.8	54.5	15.8	3.7	73.9	1.0	0.4	139.2	1.4	140.6
1994	----	64.4	56.2	15.9	3.8	75.9	1.0	0.4	141.6	1.2	142.8
1995	----	64.9	57.6	16.3	4.0	77.9	0.9	0.4	144.1	0.9	145.0

Table 2. (Continued)

Calendar year	U.S. population 2/ (July 1) Millions	Refined		Corn sweeteners			Pure honey	Edible syrups	Total caloric sweeteners	Sugar from net imports 3/	Total caloric sweeteners with sugar-containing products
		sugar	HFCS	Glucose syrup	Dextrose	Total					
1996	----	65.2	57.8	16.4	4.0	78.2	1.0	0.4	144.7	0.9	145.6
1997	----	64.9	60.4	17.3	3.7	81.5	0.9	0.4	147.8	1.4	149.2
1998	----	64.9	61.9	17.1	3.6	82.7	0.9	0.4	149.0	2.0	150.9
1999	----	66.3	63.7	16.3	3.5	83.5	1.1	0.4	151.4	2.5	153.9
2000	----	65.5	62.7	15.8	3.4	81.8	1.1	0.4	148.9	2.2	151.1
2001	----	64.5	62.6	15.5	3.3	81.4	0.9	0.4	147.3	2.7	150.0
2002	----	63.3	62.9	15.5	3.3	81.6	1.1	0.4	146.5	3.7	150.1
2003	----	61.0	61.0	15.2	3.1	79.3	1.0	0.4	141.7	4.3	146.0
2004	----	61.7	59.9	15.6	3.3	78.9	0.9	0.4	141.9	4.5	146.4
2005	----	63.2	59.2	15.3	3.3	77.8	1.1	0.4	142.5	4.5	147.0
2006	----	62.6	58.3	13.8	3.1	75.2	1.2	0.4	139.4	5.7	145.2
2007	----	62.3	56.3	13.7	3.0	73.0	0.9	0.4	136.7	5.6	142.3
2008	----	65.4	53.2	13.4	2.8	69.3	1.0	0.5	136.2	4.3	140.5
2009 4/	----	63.6	50.1	13.0	2.7	65.8	0.9	0.5	130.7	3.6	134.3

HFCS = High fructose corn syrup.

1/ Per capita deliveries of sweeteners by U.S. processors and refiners and direct-consumption imports to food manufacturers, retailers, and other end users represent the per capita supply of caloric sweeteners. The data exclude deliveries to manufacturers of alcoholic beverages. Actual human intake of caloric sweeteners is lower because of uneaten food, spoilage, and other losses.

2/ U.S. Census Bureau.

3/ Excludes sugar exported as part of USDA's Sugar-Containing Re-Export Program.

4/ Preliminary.

Source: USDA, ERS, Sugar and Sweeteners Team.

Table 3. Estimated sugar in U.S. product imports and exports, 1995-2009 5/13/10.

Year	Sugar confectionery	Cocoa and cocoa preparations	Cereal and bakers preparations	Bread, pastry, cakes, etc.	Misc. edible preparations	Carbonated soft drinks	Total sugar in imported products	Total sugar in exported products 1/	Net sugar inflow in products
1,000 short tons									
1995	137.760	66,265	6,286	43,705	68,945	26,405	349,365	317,809	31,557
1996	148.383	75,911	8,580	49,882	60,729	32,456	375,940	356,966	18,974
1997	161.894	92,664	14,273	64,812	68,172	39,403	441,218	390,159	51,060
1998	186,572	97,616	19,110	74,726	91,119	39,811	508,954	371,414	137,540
1999	223,421	111,807	20,116	87,875	118,876	48,165	610,261	392,208	218,053
2000	239,914	130,407	19,548	99,740	120,366	58,745	668,719	442,596	226,122
2001	259,975	160,350	18,097	115,917	127,331	64,961	746,630	470,991	275,639
2002	299,003	193,608	19,419	117,838	140,369	70,852	841,090	459,931	381,159
2003	362,786	208,260	25,139	134,500	150,859	83,440	964,985	507,950	457,035
2004	400,819	220,067	25,082	138,898	186,328	97,731	1,068,925	539,237	529,689
2005	456,969	231,322	26,012	143,742	187,838	109,747	1,155,630	596,960	558,670
2006	499,547	275,449	24,732	148,595	193,692	126,714	1,268,728	560,835	707,893
2007	489,598	276,990	25,081	150,538	189,345	128,811	1,260,363	588,293	672,069
2008	423,394	281,628	25,230	155,565	191,995	125,451	1,203,262	680,094	523,168
2009	398,174	264,616	20,687	154,104	181,903	121,649	1,141,132	696,951	444,182

1/ Includes sugar exported as part of USDA's Sugar-Containing Re-Export Program.

Source: USDA, ERS, Sugar and Sweeteners Team.

U.S. Sweetener Demand

The Economic Research Service (ERS) makes calendar-year estimates of total sweetener deliveries that are available for food and beverage consumption by U.S. consumers. These sweeteners include refined sugar; the corn sweeteners of HFCS, glucose syrup, and dextrose; honey; and other edible syrups, including maple syrup and maple sugar. Table 2 shows the new estimates for 2009 and also includes refined sugar revisions for 2007 and 2008.

U.S. deliveries of total sweeteners for human food and beverage use for 2009 are estimated at 20.078 million tons, representing a decrease of 3 percent compared with deliveries in 2008. Refined sugar deliveries decreased by 1.8 percent, and corn sweetener deliveries fell by 4.1 percent. Within the corn sweetener category, HFCS deliveries fell by 4.7 percent. Since 2002, HFCS deliveries have fallen by 1.35 million tons, dry weight, or 14.9 percent. Honey deliveries decreased by 6.7 percent, and other edible syrups increased slightly.

U.S. sweetener deliveries for 2009 were equal to 130.7 pounds per capita, down 5.5 pounds from 2008 and down 20.7 pounds from the per capita high of 151.4 pounds set in 1999. About 65 percent of the decline from 2008 is attributable to lower corn sweetener deliveries and about 34 percent from lower refined sugar deliveries. However, per capita sugar consumption in 2009 at 63.6 pounds is still higher than its level of 62.3 pounds 2 years ago.

Sweeteners at-a-Glance

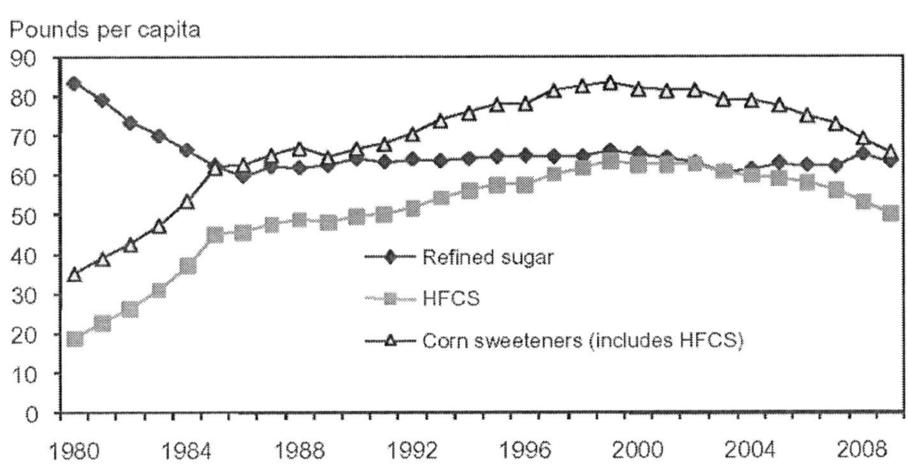

Source: USDA, ERS, Sugar and Sweeteners Team.

Figure A. Per capita sweetener availability: refined sugar, high fructose corn syrup (HFCS), all corn sweeteners, 1980-2009.

Sugar contained in imported products is usually excluded in estimating U.S. per capita sweetener deliveries. Before 1995, sugar contained in imports was offset by sugar contained in U.S. food exports, therefore indicating only a minor positive adjustment to total deliveries. Beginning in the 1995-96 period, imports of sugar-containing products started increasing at a faster rate than U.S. exports of sugar-containing products. This trend continued until 2006 and has since reversed. For 2009, sugar-containing product trade has contributed an estimated

555,000 tons to sweetener available for consumption, or 3.6 pounds per capita. This level is down from the high of 5.7 pounds in 2006.

Sugar in imported products in 2009 is estimated at 1.141 million tons, a 5.2-percent decrease from 2008 and 10.1- percent decrease from 2006 (table 3). About 41 percent of the decrease since 2008 is attributable to fewer imports of sugar confectionery, and about 27 percent is attributable to fewer imports of cocoa and cocoa preparations.

Sugar in exported products has been growing since 2006, estimated at 697,000 tons in 2009, up about 2.5 percent over 2008 and 24.3 percent over 2006.

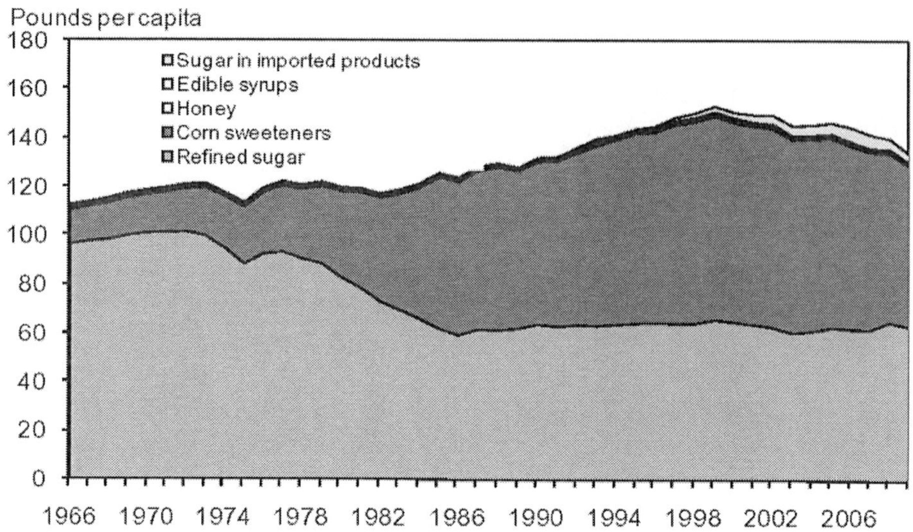

Source: USDA, ERS, Sugar and Sweeteners Team.

Figure B. U.S. per capita sweetener consumption, 1966-2009.

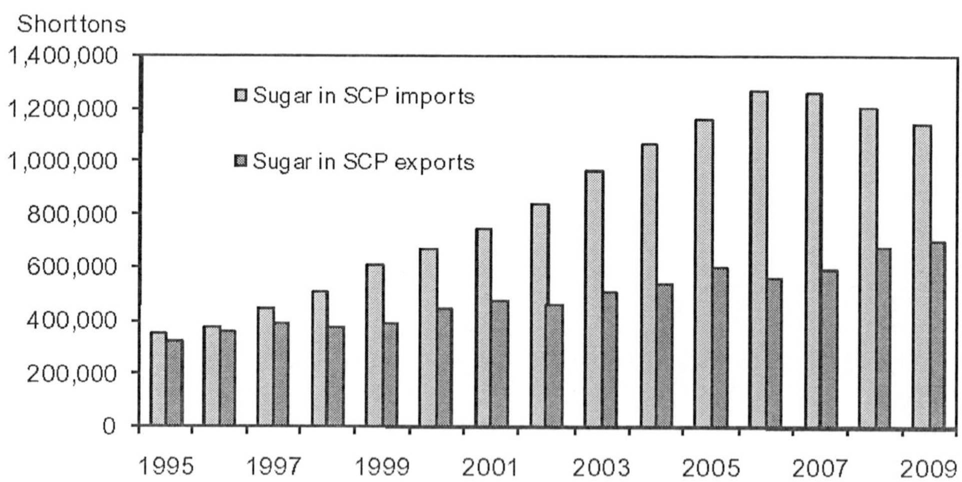

Source: USDA, ERS, Sugar and Sweeteners Team.

Figure C. Estimated sugar contained in U.S. imports and exports of certain sugar-containing products (SCP), 1995-2009.

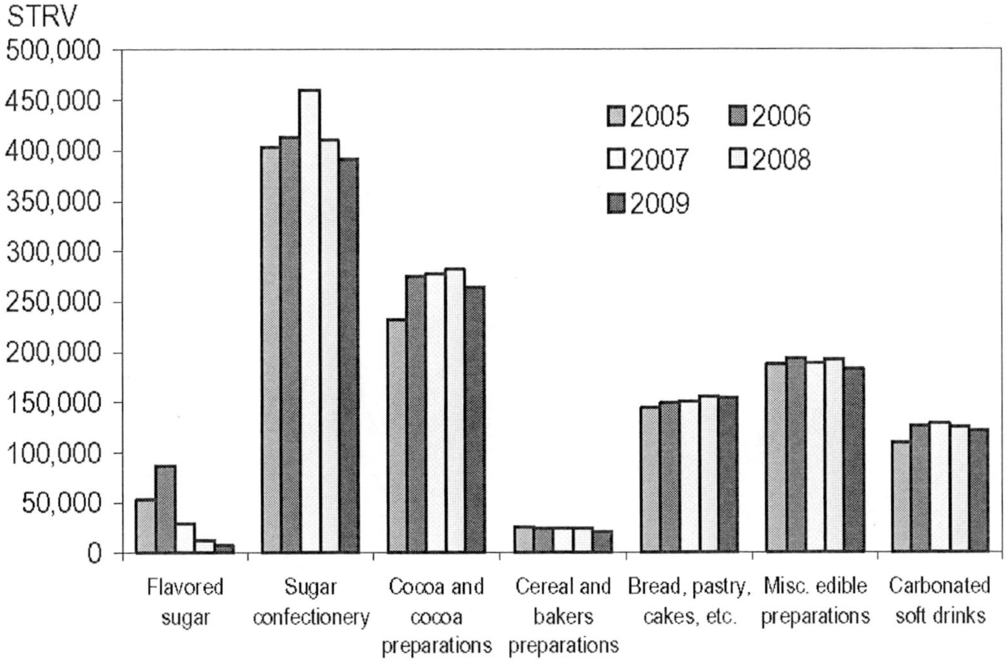

Source: USDA, ERS, Sugar and Sweeteners Team.

Figure D. Sugar in product imports, by product category, 2005-09.

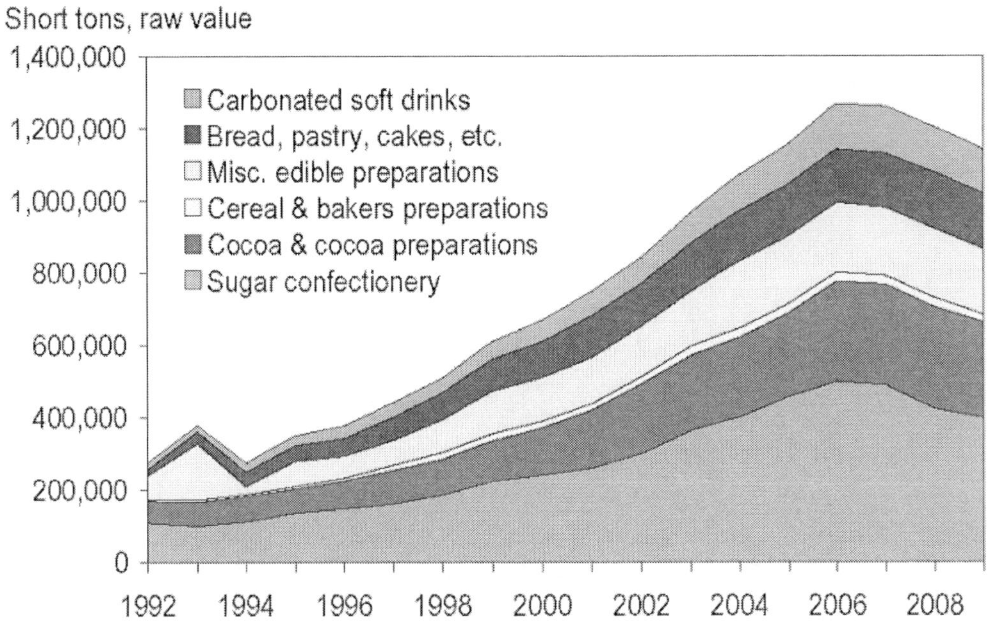

Source: USDA, ERS, Sugar and Sweeteners Team.

Figure E. Sugar contained in certain sugar-containing product imports.

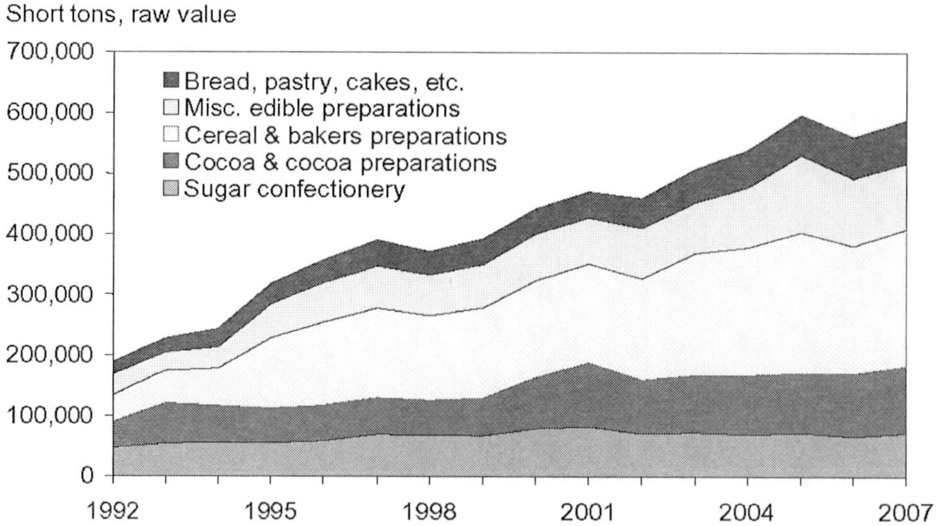

Source: USDA, ERS, Sugar and Sweeteners Team.

Figure F. Sugar contained in certain sugar-containing product exports.

MEXICAN SUGAR & HIGH FRUCTOSE CORN SYRUP

Table 4. U.S. sugar: Supply and use, by fiscal year, short tons 1/ 5/13/2010.

Items	2003/ 04	2004/ 05	2005/ 06	2006/ 07	2007/ 08	2008/ 09	2009/ 10	201 0/11 Projection 10-May
1,000 short tons, raw value								
Beginning stocks 2/	1,670	1,897	1,332	1,698	1,799	1,660	1,499	1,230
Total production 3/,4/	8,649	7,876	7,399	8,445	8,152	7,532	7,847	8,165
Beet sugar	4,692	4,611	4,444	5,008	4,721	4,214	4,500	4,630
Cane sugar	3,957	3,265	2,955	3,438	3,431	3,318	3,347	3,535
Florida	2,154	1,693	1,367	1,719	1,645	1,577	1,630	1,785
Louisiana	1,377	1,157	1,190	1,320	1,446	1,397	1,465	1,465
Texas	175	158	175	177	158	152	116	150
Hawaii	251	258	223	222	182	192	136	135
Puerto Rico	0	0	0	0	0	0	0	
Total imports	1,750	2,100	3,443	2,080	2,620	3,082	2,519	2,084
Tariff-rate quota imports 5/	1,226	1,408	2,588	1,624	1,354	1,370	1,554	1,224
Other program imports	464	500	349	390	565	308	350	300
Nonprogram imports	60	192	506	66	701	1,404	615	560
Mexico 6/				60	694	1,402	540	550
Total supply	12,070	11,873	12,174	12,223	12,571	12,274	11,865	11,479

Table 4. (Continued)

Items	2003/ 04	2004/ 05	2005/ 06	2006/ 07	2007/ 08	2008/ 09	2009/ 10	201 0/11 Projection 10-May
Total exports 3/	288	259	203	422	203	137	150	150
Quota-exempt for reexport	288	259	203	422	203	137	150	150
Other exports	0	0	0					
CCC disposal, for export	0	0	0					
Miscellaneous	23	94	-67	-132	0	0	0	0
CCC disposal, for domestic nonfood use	0	0	0	0	0	0	0	0
Refining loss adjustment	0	0	0	0	0	0	0	0
Statistical adjustment 7/	23	94	-67	-132	0	0	0	0
Deliveries for domestic use	9,862	10,188	10,340	10,135	10,708	10,638	10,485	10,485
Transfer to sugar-containing products for exports under re-export program	142	121	106	169	141	113	145	145
Transfer to polyhydric alcohol, feed	41	48	51	53	61	46	40	40
Deliveries for domestic food and beverage use 8/	9,678	10,019	10,184	9,913	10,506	10,479	10,300	10,300
Total use	10,172	10,542	10,476	10,424	10,912	10,775	10,635	10,635
Ending stocks 2/	1,897	1,332	1,698	1,799	1,660	1,499	1,230	844
Privately owned CCC								
				Percent				
Stocks-to-use ratio	18.65	12.63	16.21	17.25	15.22	13.91	11.57	7.94

CCC = Commodity Credit Corporation.

Note: Numbers may not add due to rounding.

1/ Fiscal year beginning October 1. 2/ Stocks in hands of primary distributors and CCC. 3/ Historical data are from Farm Service Agency *(FSA) for Sweetener Market Data* (SMD) and USDA, National Agricultural Statistics Service, *Sugar Market Statistics* prior to 1992.

4/ Production reflects processors' projection compiled by FSA. 5/ Actual arrivals under the tariff-rate quota (TRQ) with late entries, early entries, and TRQ overfills assigned to the fiscal year in which they actually arrived. The 2009/10 available TRQ assumes shortfall of 200,000 tons. 6/ Starting in 2007/08, total includes imports under Mexico's World Trade Organization TRQ allocation for raw and refined sugar. 7/ Calculated as a residual. Largely consists of invisible stocks change. 8/ For FY 2008-10, combines SMD deliveries for domestic human use, and the difference between SMD imports and *World Agricultural Supply and Demand Estimates* imports.

Table 5. U.S. sugar: supply and use (including Puerto Rico), fiscal years, metric tons 1/, 5/13/2010.

Items	2003/04	2004/05	2005/06	2006/07	2007/08	2008/09	2009/10	2010/11
1,000 metric tons, raw value								
Beginning stocks 2/	1,515	1,721	1,208	1,540	1,632	1,506	1,360	1,116
Total production 3/ 4/	7,846	7,145	6,712	7,662	7,396	6,833	7,119	7,407
Beet sugar	4,257	4,183	4,032	4,543	4,283	3,823	4,082	4,200
Cane sugar	3,590	2,962	2,681	3,119	3,113	3,010	3,036	3,207
Florida	1,954	1,536	1,241	1,559	1,492	1,431	1,479	1,619
Louisiana	1,249	1,049	1,079	1,198	1,312	1,267	1,329	1,329
Texas	159	143	159	161	143	138	105	136
Hawaii	228	234	202	201	165	174	123	122
Puerto Rico	0	0	0	0	0	0	0	0
Total imports	1,588	1,905	3,124	1,887	2,377	2,796	2,285	1,891
Tariff-rate quota imports 5/	1,113	1,277	2,348	1,473	1,228	1,243	1,410	1,110
Other program imports	421	454	317	354	513	279	318	272
Nonprogram imports	54	174	459	60	636	1,274	558	508
Mexico 6/	0	0	0	54	630	1,272	490	499
Total supply	10,950	10,771	11,044	11,088	11,404	11,135	10,764	10,414
Total exports 3/	261	235	184	383	184	124	136	136
Quota-exempt for re-export	261	235	184	383	184	124	136	136
Other exports	0	0	0	0	0	0	0	0
CCC disposal, for export	0	0	0	0	0	0	0	0
Miscellaneous	20	85	-61	-120	-1	1	0	0
CCC disposal, for domestic nonfood use	0	0	0	0	0	0	0	0
Refining loss adjustment	0	0	0	0	0	0	0	0
Statistical adjustment 7/	20	85	-61	-120	-1	1	0	0
Deliveries for domestic use	8,947	9,243	9,381	9,194	9,715	9,650	9,512	9,512
Transfer to sugar-containing products for exports under re-export program	129	110	96	153	128	102	132	132
Transfer to polyhydric alcohol, feed	38	44	46	48	56	42	36	36
Deliveries for domestic food and beverage use 8/	8,780	9,089	9,239	8,993	9,531	9,506	9,344	9,344
Total use	9,228	9,563	9,504	9,457	9,898	9,775	9,648	9,648
Ending stocks 2/	1,721	1,208	1,540	1,632	1,506	1,360	1,116	766
Privately owned	0	0	0	0	0	0	0	0
CCC	0	0	0	0	0	0	0	0
Percent								
Stocks-to-use ratio	18.65	12.63	16.21	17.25	15.22	13.91	11.57	7.94

CCC = Commodity Credit Corporation.

Note: Numbers may not add due to rounding.

1/ Fiscal year beginning October 1. 2/ Stocks in hands of primary distributors and CCC. 3/ Historical data are from USDA, Farm Service Agency (FSA) (formerly ASCS), *Sweetener Market Data* (SMD) and USDA, NASS, *Sugar Market Statistics* prior to 1992. 4/ Production reflects processors' projections compiled by the FSA .

5/ Actual arrivals under the tariff-rate quota (TRQ) with late entries, early entries, and TRQ overfills assigned to the fiscal year in which they actually arrived. The 2009/10 available TRQ assumes shortfall of 181,437 tons. 6/ Starting in 2007/08, total includes imports under Mexico's World Trade Organization TRQ allocation for raw and refined sugar. 7/ Calculated as a residual. Largely consists of invisible stocks change. 8/ For FY 2008-10, combines SMD deliveries for domestic and human use, SMD miscellaneous uses, and the difference between SMD imports and *World Agricultural Supply and Demand Estimates* imports.

Mexican sugar production is forecast to be 5.45 million metric tons, raw value (MTRV) in the 2010/11 harvest season, an 11-percent increase from the previous year. Expanded harvest area and recovering sugarcane yields will contribute to higher production. Area and yields were lower in 2008/09 and 2009/10 compared with historical trends, primarily because of poor weather conditions, contributing to lower production levels.

Higher U.S. and Mexican sugar prices in 2009/10 are expected to encourage planted and harvested area for sugarcane in 2010/11. Harvested area in the 2008/09 and 2009/10 growing seasons were 663,000 and 664,000 hectares, respectively, a 3-percent decrease from the 683,000 hectares harvested in 2007/08. The decrease was primarily due to poor weather conditions. In 2010/11, harvested area is expected to increase, although it is not expected to fully recover to 2007/08 levels.

Weather has also negatively affected sugarcane yields the past two seasons. From 1989/90 through 2008/09, sugarcane yields averaged 72 metric tons per hectare. Yields in 2008/09 and 2009/10 were, or are expected to be, nearly 12 percent lower than that of the 20-year average. Assuming normal weather conditions, yields are anticipated to rebound in 2010/11 and will be more aligned with average yields.

Sucrose recovery rates in Mexico have demonstrated a statistically significant upward trend over the past 20 years. The trend implies that recovery rates in 2010/11 should be approximately 11.4 percent, which was the basis for the estimate in the *World Agriculture Supply and Demand Estimates*. This rate is above the 2009/10 pace to date of 11 percent but not quite as high as the 2007/08 recovery rate of 11.7 percent, which was a 20-year high.

Domestic sugar use is expected to fall slightly from 5 million MTRV in 2009/10 to 4.95 million MTRV in 2010/11. Domestic sugar and sweetener demand for human consumption is forecast to increase 3 percent based on increasing per capita consumption patterns. However, HFCS accounts for all of the increase. Human domestic sugar consumption is expected to decrease 1 percent from 4.6 million metric tons to 4.55 million metric tons. The IMMEX re-export program is anticipated to be fully used in the upcoming year, accounting for the remaining 400,000 metric tons of domestic use.

Mexican sugar imports are forecast to be 150,000 MTRV compared with the estimate for 2009/10 of 970,000 MTRV. With improved production and higher beginning stocks, imports will be primarily from the 135,000 metric tons allotted to the IMMEX re-export program.

Mexican sugar exports are expected to increase slightly next year from 490,000 metric tons to 500,000 metric tons, primarily going to the United States. Higher returns and a relatively tighter supply situation in the United States should continue to encourage Mexican sugar to be marketed in the United States. Since the beginning of the 2008 calendar year, Mexico has been able to export sugar to the United States without quota restrictions as part of the North American Free Trade Agreement.

Ending stocks for 2010/11 are expected to be 1.018 million tons, which would imply a stocks-to-domesticconsumption ratio of 22 percent compared with the current estimate of 19 percent in 2009/10. Higher production levels should help alleviate the tight stock levels that were observed over the previous 2 marketing years.

Table 6. Mexico: Sugar production and supply and sugar and high fructose corn syrup (HFCS) utilization, by fiscal years 5/13/2010.

Fiscal year (Oct/Sept)	2003	2004	2005	2006	2007	2008	2009	2010 1/	2011 1/
				1,000 metric tons					
Beginning stocks	1,172	1,194	1,237	1,965	1,294	1,718	1,975	488	868
Production	5,229	5,330	6,149	5,604	5,633	5,852	5,260	4,900	5,450
Imports	63	327	268	240	474	226	160	970	150
Supply	6,464	6,851	7,654	7,809	7,401	7,796	7,395	6,358	6,468
Disappearance									
Human consumption	5,097	5,380	5,279	5,326	5,133	5,090	5,065	4,600	4,550
Other consumption	135	220	282	323	390	414	475	400	400
Miscellaneous						-360			
Total	5,232	5,600	5,561	5,649	5,523	5,144	5,540	5,000	4,950
Exports	38	14	128	866	160	677	1,367	490	500
Total use	5,270	5,614	5,689	6,515	5,683	5,821	6,907	5,490	5,450
Ending stocks	1,194	1,237	1,965	1,294	1,718	1,975	488	868	1,018
Stocks-to-human consumption	23.4	23.0	37.2	24.3	33.5	38.8	9.6	18.9	22.4
Stocks-to-use	22.7	22.0	34.6	19.9	30.2	33.9	7.1	15.8	18.7
HFCS consumption (dry weight)	130	135	355	667	698	782	653	1,200	1,400

1/ Forecast.

Sources: USDA, Foreign Agricultural Service, Production, Supply and Distribution database (historical data); *World Agricultural Supply and Demand Estimates.*

Contacts and Links

Data

Tables from the *Sugar and Sweeteners Yearbook* are available in the Sugar and Sweeteners Briefing Room at http://www.ers.usda.gov/briefing/sugar/. They contain the latest data and historical information on the production, use, prices, imports, and exports of sugar and sweeteners.

Related Websites

Sugar and Sweeteners Outlook http://www.ers.usda.gov/Publications/SSS/

WASDE
http://usda.mannlib.cornell.edu/MannUsda/viewDocumentInfo.do?documented=1194
Sugar Briefing Room, http://www.ers.usda.gov/briefing/Sugar/

E-mail Notification

Readers of ERS outlook reports have two ways they can receive an e-mail notice about release of reports and associated data.

- Receive timely notification (soon after the report is posted on the web) via USDA's Economics, Statistics and Market Information System (which is housed at Cornell University's Mann Library). Go to http://usda.mannlib.cornell.edu/MannUsda/aboutEmailService.do and follow the instructions to receive e-mail notices about ERS, Agricultural Marketing Service, National Agricultural Statistics Service, and World Agricultural Outlook Board products.
- Receive weekly notification (on Friday afternoon) via the ERS website. Go to http://www.ers.usda.gov/Updates/ and follow the instructions to receive notices about ERS outlook reports, Amber Waves magazine, and other reports and data products on specific topics. ERS also offers RSS (really simple syndication) feeds for all ERS products. Go to http://www.ers.usda.gov/rss/ to get started.

The U.S. Department of Agriculture (USDA) prohibits discrimination in all its programs and activities on the basis of race, color, national origin, age, disability, and, where applicable, sex, marital status, familial status, parental status, religion, sexual orientation, genetic information, political beliefs, reprisal, or because all or a part of an individual's income is derived from any public assistance program. (Not all prohibited bases apply to all programs.) Persons with disabilities who require alternative means for communication of program information (Braille, large print, audiotape, etc.) should contact USDA's TARGET Center at (202) 720-2600 (voice and TDD). To file a complaint of discrimination write to USDA, Director, Office of Civil Rights, 1400 Independence Avenue, S.W., Washington, D.C. 20250-9410 or call (800) 795-3272 (voice) or (202) 720-6382 (TDD). USDA is an equal opportunity provider and employer.

End Notes

[1] A closely related forecast model takes account of sucrose content and produces better statistical results. However, NASS does not release sucrose content until late into the crop year, meaning that the level for 2009/10 is not yet available.

In: Sugar: Background, Policies and Issues
Editor: Cecilia M. Jordan

ISBN: 978-1-61761-116-2

Chapter 6

SUGAR AND SWEETENERS OUTLOOK (FEBRUARY 2010 SUGAR)*

Stephen Haley

In the February 2010 *World Agriculture Supply and Demand Estimates* (WASDE), projected fiscal year (FY) 2010 production for Mexico is reduced 200,000 metric tons, raw value (MTRV) from last month based on weather-reduced sugar yields to date. Exports are reduced by the same amount. Projected FY 2010 U.S. sugar supply is decreased by 85,000 short tons, raw value (STRV) from last month due to lower imports from Mexico, more than offsetting higher sugar production. Imports from Mexico are reduced by 220,000 STRV. Beet sugar production and Louisiana cane sugar production are increased a combined 150,000 STRV due to anticipated output in September 2010 from next season's sugar crops. Florida cane sugar production is reduced 15,000 STRV based on a lower forecast of sugarcane for this year by the National Agricultural Statistics Service (NASS). Total use is unchanged. Ending-year stocks are projected at 1.055 million STRV, implying an ending stocks-to-use ratio of 10.0 percent.

U.S. Sugar

In the February 2010 *World Agricultural Supply and Demand Estimates* (WASDE), the U.S. Department of Agriculture (USDA) published its latest projections for fiscal year (FY) 2010. Expected U.S. beet sugar production is increased by 100,000 short tons, raw value (STRV) and is now projected at 4.500 million STRV. The increase is due to expected production in September from next season's sugar beet harvest and is not attributable to this year's harvest (see below for a summary of the 2009/10 sugar beet harvest).

The cane sugar forecast for FY 2010 is 3.472 million STRV, an increase of 35,000 STRV from last month's projection. The forecast for Louisiana is increased by 50,000 STRV to

* This is an edited, reformatted and augmented edition of a United States Department of Agriculture publication, Report SSS-M-258, dated February 11, 2010.

1.500 million STRV. Similar to the case for beet sugar, this increase stems from expected September production from next season's sugarcane harvest. In the February 2010 Crop Production, the National Agricultural Statistics Service (NASS) lowered its projection of Florida sugarcane for sugar by 0.8 percent to 13.429 million tons, noting that freezing temperatures in mid-January had lowered expected yield by 0.3 tons to 36.1 tons/acre. Cane sugar production from Florida is now forecast at 1.665 million STRV.

Projected sugar production in Texas and Hawaii were left unchanged at a combined total of 307,000 STRV. Nonetheless, cold and wet conditions in Texas left December sugar production at only 1,259 STRV. Normally a total for December Texas production would be between 28,000 and 32,000 STRV.

Due to lower-than-expected sugar production in Mexico (see below), sugar imports from Mexico are reduced by 220,000 STRV and are now projected at 540,000 STRV. Mexican imports for the October-December period have totaled 105,820 STRV. All sugar has entered by surface transport and most is likely intended for direct consumption—that is, not for refining.

Total sugar imports through the end of December (not counting high-tier tariff sugar) are estimated at 701,000 STRV, or about 33.0 percent of expected imports. Raw sugar tariff-rate quota (TRQ) entries through February 2 are estimated at a relatively high amount of 558,910 STRV, or 54.0 percent of the expected total through 4 months of the fiscal year.

Total use projections were left unchanged. Exports are projected at 150,000 STRV. These are mostly exports occurring under the Refined Sugar Re-Export Program. The sum of deliveries for human use and miscellaneous use are projected at 10.140 million STRV. Other deliveries are projected at 235,000 STRV. These are the sum of deliveries for the Sugar-Containing Products Program, the polyhydric alcohol program, and livestock feed. Total ending-year stocks are projected at 1.055 million STRV, implying an ending-year stocks-to-use ratio of 10 percent.

In the *Sweetener Market Data* (SMD), deliveries for October-December 2009 are estimated at 2.73 6 million STRV, or about 4.3 percent more than deliveries for the same corresponding period last year. However, the Sugar and Sweetener Team of the Economic Research Service (ERS) believes that the SMD estimate of direct consumption imports is overstated by 194,000 STRV. Although SMD errors should be self-correcting over time, monthly estimates are deemed unreliable. The ERS Team estimates deliveries for human consumption through December 2009 at 2.591 million STRV, or 4.5 percent below the same period last year. More detailed analysis will be provided in the next issue of the *Sugar and Sweeteners Outlook*.

The Inter-Continental Exchange (ICE) raw sugar March No.16 futures contract has averaged 36.5 cents/pound (lb) the first week and a half of February, more than 12 cents higher than the corresponding world price (No.11 ICE March contract). The high margin reflects tightness in domestic supplies and is needed to keep the U.S. competitive in bidding TRQ sugar away from other importing countries. Also, uncertainty about future TRQ increases and Mexican imports are being factored into the market. In like fashion, the refined beet sugar price in the Midwest has been quoted by *Milling and Baking News* at 53 cents/lb since the week ending January 22.

Challenging weather conditions affected most sugar beet producers in 2009. Most areas experienced wet conditions during planting and harvest seasons. The cool, dry summer adversely limited yield potential during the growing season. However, only limited reports of root disease and pest issues, partially due to increased use of Roundup Ready and disease-resistant varieties of beets, offset weather-related obstacles and resulted in relatively strong yields according to many of the largest sugar beet cooperatives. Sugar content in the crop though was typically lower than average, particularly in the Upper Midwest, the result of the cool summer and early fall freezes.

Sugar beet producers in the Upper Midwest contended with wet conditions during both planting and harvest, delaying both activities. Producers near the Red River Valley, which was affected by severe flooding in the spring, reported that plantings were delayed over 2 weeks and some cases of abandoned acreage due to the wet conditions. A dry, cool summer affected root growth development in some areas, but root disease and fungus issues were not widely reported, resulting in strong yields by the larger cooperatives in Minnesota and North Dakota. Wet conditions returned in September, in time to complicate or stall the harvest. High levels of soil tare were reported in the region. Early freezes in September and October also pushed the harvest later into the fall and kept sugar content levels relatively low.

Further west, in Idaho, Wyoming, Nebraska, Montana, and Colorado, planting conditions were generally more favorable. Some areas that experienced drought conditions in prior years saw more favorable water levels in 2009. There were some reports of delayed plantings and replantings in Idaho due to April snow and in Montana due to a late freeze in May. Overall, these States increased plantings over 20 percent in 2009 compared with the previous year, according to NASS. The cool summer limited the sugar content of the roots, but no major reports of pest or disease issues kept yields relatively high in these areas. Like other regions, a wet fall pushed the harvest later into the fall and freezing temperatures further limited sugar content on the crop.

Michigan sugar beet production shrank in 2009 from that of 2008. Heavy rains during the planting period caused seedling disease and soil crusting. Lack of precipitation during the growing season was also detrimental to overall yields. Additionally, rhizoctonia crown rot was prevalent in the region according to a cooperative in the area, even for resistant varieties of beets. A dry fall helped with a quick harvest, and cool temperatures in December made for good storage conditions.

Finally, California's acreage planted was lower in 2009 due to delays in contract negotiations and pressure from alternative crops. High temperatures raised pressures from pests and threatened yields; however, improving conditions and a warm winter helped offset some of the decreases in acreage with a higher yield, including a record high yield for the Imperial Valley in Southern California.

Overall, the 2009 sugar beet season was marked by difficult planting and harvesting conditions. A wet spring and cool summer limited root growth, but advanced seed varieties helped to manage weeds and combat root disease, allowing for strong final yields. Freezing temperatures early in the harvest season limited sugar content, with northern growers reporting sugar content levels between 15.5 percent and 17 percent, as opposed to historical averages, which are closer to 17-18 percent. Total sugar beet production in 2009 was 29.5 million tons, nearly a 10-percent increase from that of 2008, according to NASS's January *Crop Production Summary*. However, sugar beet production continued to trend downward during the past 4 years.

Table 1. U.S. sugar: Supply and use, by fiscal year, short tons 1/ 2/11/2010.

Items	2003/04	2004/05	2005/06	2006/07	2007/08	2008/09	2009/10
			1,000 short tons, raw value				
Beginning stocks 2/	1,670	1,897	1,332	1,698	1,799	1,660	1,451
Total production 3/,4/	8,649	7,876	7,399	8,445	8,152	7,484	7,972
Beet sugar	4,692	4,611	4,444	5,008	4,721	4,166	4,500
Cane sugar	3,957	3,265	2,955	3,438	3,431	3,318	3,472
Florida	2,154	1,693	1,367	1,719	1,645	1,577	1,665
Louisiana	1,377	1,157	1,190	1,320	1,446	1,397	1,500
Texas	175	158	175	177	158	152	170
Hawaii	251	258	223	222	182	192	137
Puerto Rico	0	0	0	0	0	0	0
Total imports	1,750	2,100	3,443	2,080	2,620	3,082	2,157
Tariff-rate quota imports 5/	1,226	1,408	2,588	1,624	1,354	1,370	1,257
Other program imports	464	500	349	390	565	308	350
Nonprogram imports	60	192	506	66	701	1,404	550
Mexico 6/				60	694	1,402	540
Total supply	12,070	11,873	12,174	12,223	12,571	12,226	11,580
Total exports 3/	288	259	203	422	203	137	150
Quota-exempt for reexport	288	259	203	422	203	137	150
Other exports	0	0	0				
CCC disposal, for export	0	0	0				
Miscellaneous	23	94	-67	-132	0	0	0
CCC disposal, for domestic nonfood use	0	0	0	0	0	0	0
Refining loss adjustment	0	0	0	0	0	0	0
Statistical adjustment 7/	23	94	-67	-132	0	0	0
Deliveries for domestic use	9,862	10,188	10,340	10,135	10,708	10,638	10,375
Transfer to sugar-containing products for exports under reexport program	142	121	106	169	141	113	175
Transfer to polyhydric alcohol, feed	41	48	51	53	61	46	60
Deliveries for domestic food and beverage use 8/	9,678	10,019	10,184	9,913	10,506	10,479	10,140
Total use	10,172	10,542	10,476	10,424	10,912	10,775	10,525
Ending stocks 2/	1,897	1,332	1,698	1,799	1,660	1,451	1,055
Privately owned							
CCC							
			Percent				
Stocks-to-use ratio	18.65	12.63	16.21	17.25	15.22	13.47	10.02

CCC = Commodity Credit Corporation.

Note: Numbers may not add due to rounding.

1/ Fiscal year beginning October 1. 2/ Stocks in hands of primary distributors and CCC. 3/ Historical data are from Farm Service Agency (FSA) (formerly ASCS) Sweetener Market Data (SMD), and NASS, Sugar Market Statistics prior to 1992. 4/ Production reflects processors' projections compiled by the FSA.

5/ Actual arrivals under the tariff-rate quota (TRQ) with late entries, early entries, and TRQ overfills assigned to the fiscal year in which they actually arrived. The 2009/10 available TRQ assumes shortfall of 200,000 tons. 6/ Starting in 2007/08, total includes imports under Mexico's WTO TRQ allocation for raw and refined sugar. 7/ Calculated as a residual. Largely consists of invisible stocks change. 8/ For FY 2008-10, combines SMD deliveries for domestic human use, SMD miscellaneous uses, and the difference between SMD imports and WASDE imports.

Table 2. U.S. sugar: supply and use (including Puerto Rico), fiscal years, metric tons 1/ 2/11/2010.

Items	2003/04	2004/05	2005/06	2006/07	2007/08	2008/09	2009/10
	1,000 metric tons, raw value						
Beginning stocks 2/	1,515	1,721	1,208	1,540	1,632	1,506	1,316
Total production 3/ 4/	7,846	7,145	6,712	7,662	7,396	6,789	7,232
Beet sugar	4,257	4,183	4,032	4,543	4,283	3,779	4,082
Cane sugar	3,590	2,962	2,681	3,119	3,113	3,010	3,150
Florida	1,954	1,536	1,241	1,559	1,492	1,431	1,510
Louisiana	1,249	1,049	1,079	1,198	1,312	1,267	1,361
Texas	159	143	159	161	143	138	154
Hawaii	228	234	202	201	165	174	124
Puerto Rico	0	0	0	0	0	0	0
Total imports	1,588	1,905	3,124	1,887	2,377	2,796	1,957
Tariff-rate quota imports 5/	1,113	1,277	2,348	1,473	1,228	1,243	1,140
Other program imports	421	454	317	354	513	279	318
Nonprogram imports	54	174	459	60	636	1,274	499
Mexico 6/	0	0	0	54	630	1,272	490
Total supply	10,950	10,771	11,044	11,088	11,404	11,092	10,505
Total exports 3/	261	235	184	383	184	124	136
Quota-exempt for re-export	261	0	184	383	184	124	136
Other exports	0	0	0	0	0	0	0
CCC disposal, for export	0	235	0	0	0	0	0
Miscellaneous	20	85	-61	-120	-1	1	0
CCC disposal, for domestic nonfood use	0	0	0	0	0	0	0
Refining loss adjustment	0	0	0	0	0	0	0
Statistical adjustment 7/	20	85	-61	-120	-1	1	0
Deliveries for domestic use	8,947	9,243	9,381	9,194	9,715	9,650	9,412
Transfer to sugar-containing products for exports under reexport program	129	110	96	153	128	102	159
Transfer to polyhydric alcohol, feed	38	44	46	48	56	42	54
Deliveries for domestic food and beverage use 8/	8,780	9,089	9,239	8,993	9,531	9,506	9,199
Total use	9,228	9,563	9,504	9,457	9,898	9,775	9,548
Ending stocks 2/	1,721	1,208	1,540	1,632	1,506	1,316	957
Privately owned	0	0	0	0	0	0	0
CCC	0	0	0	0	0	0	0
	Percent						
Stocks-to-use ratio	18.65	12.63	16.21	17.25	15.22	13.47	10.02

CCC = Commodity Credit Corporation.

Note: Numbers may not add due to rounding.

1/ Fiscal year beginning October 1. 2/ Stocks in hands of primary distributors and CCC. 3/ Historical data are from The Farm Service Agency (FSA) (formerly ASCS), Sweetener Market Data (SMD), and NASS, Sugar Market Statistics prior to 1992. 4/ Production reflects processors' projections compiled by the FSA .

5/ Actual arrivals under the tariff-rate quota (TRQ) with late entries, early entries, and TRQ overfills assigned to the fiscal year in which they actually arrived. The 2009/10 available TRQ assumes shortfall of 181,437 tons. 6/ Starting in 2007/08, total includes imports under Mexico's WTO TRQ allocation for raw and refined sugar. 7/ Calculated as a residual. Largely consists of invisible stocks change. 8/ For FY 2008-10, combines SMD deliveries for domestic human use, SMD miscellaneous uses, and the difference between SMD imports and WASDE imports.

Mexico Sugar and HFCS

Mexico sugar production for fiscal year (FY) 2010 is forecast at 5.100 million metric tons, raw value (MTRV), a reduction of 200,000 MTRV from the forecast last month. Although more area has been harvested this year through January 30 compared with last year (220,719 hectares (ha) compared with 184,929 ha) and more sugarcane harvested (14,158,544 metric tons (mt) compared with 13,870,989 mt), sugar produced is 4.1 percent lower (1.450 million mt compared with 1.512 million mt, actual weight). These data indicate sugarcane yield at 64.15 mt/ha, 14.5 percent lower than last year, and sugar yield at 6.569 mt/ha, 19.7 percent lower. The harvest has been affected by unusually cool, damp harvest conditions, especially in Veracruz, the largest sugar-producing State in Mexico. Production there is 10.2 percent lower than the same corresponding period last year. The weak start is considered an indicator of the poor prospects through the rest of the campaign. Other groups making production forecasts, like the National Sugarcane Growers Union, have reduced their forecasts recently.

The USDA projects Mexican sugar exports at 490,000 MTRV, a 200,000-MTRV reduction paralleling the lower production forecast. Sugar prices, which had been declining through the end of 2009 because of sugar imports, increased in January. The Mexico City estandar price began the month at 37.6 cents/pound (lb) and finished at 43.8 cents/lb. The Mexico City refined price began at 41.4 cents/lb and was above 48 cents/lb midmonth, averaging 45.9 cents the entire month. These increased prices plus policymaker concerns about the affordability of domestic sugar are factors arguing for fewer exports. Even so, the export projection has a wide band around it. U.S. refined prices, Midwest, are higher in Mexico City at 53 cents/lb, but U.S. sugar raw prices are still lower. Also, there are reports that the Mexican Government's Secretariat of the Economy will officially announce a tariff-rate quota of up to 250,000 mt, providing for entry through mid-May.

The USDA projects Mexican consumption of high fructose corn syrup (HFCS) at 900,000 mt, dry weight, the same as last month. Depending on Mexican HFCS production, Mexican HFCS imports are expected to be between 450,000-550,000 mt, most of which are expected to come from the United States. For the first 2 months of the fiscal year (October and November), the United States has shipped to Mexico 136,500 mt of HFCS, more than 2.5 times as much as in the same period the previous year. This would be between 25-30 percent of the expected total. Exports for December 2009 will be released soon after the *World Agricultural Supply and Demand Estimates* and will likely give additional guidance for next month's forecast.

Mexican Sugar and HFCS Long-Term Projections through Fiscal Year 2021

The USDA prepares long-term sugar projections for both the United States and Mexico in the fall for publication prior to the Agricultural Outlook Forum in February of the following year. First-year projections (2009/10 October/September marketing year—same as the U.S. fiscal year (FY) 2010) are the same as those published by the USDA's Foreign Agricultural Service (FAS) in the November 2009 Production, Supply, and Distribution (PSD) database. Mexican production for FY 2010 is forecast at 5.400 million metric tons, raw value (MTRV). (The forecast is lower than that now—see previous section of this chapter.) Consumption for food and beverages is forecast at 4.900 million MTRV, which is 165,000 MTRV lower than the previous year. High domestic sugar prices are expected to lead to increased consumption of high fructose corn syrup (HFCS), forecast at 900,000 metric tons (mt), dry basis.[1] This is 290,000 mt greater than the estimated 610,000 mt in FY 2009. It is assumed that the beverage industry consumes about 80 percent of that amount. HFCS is estimated to make up about 48 percent of combined sugar and HFCS sweeteners demanded by the Mexican beverage industry. Sugar exports are forecast at 690,000 MTRV, and almost all exported sugar is expected to be shipped to the United States. Ending stocks are forecast at 760,000 MTRV, implying a stocks-to-human consumption ratio of 15.5 percent. This stock level is low by historical standards. This chapter assumes that policymakers typically aim for a stocks ratio of about 22 percent to ensure adequate supplies from the end of the fiscal year to the beginning of the next season's sugarcane harvest.

Projections framework

The Mexico sugar and sweetener projections are determined simultaneously with those for the United States. (The U.S. projections are part of a separate document released by the USDA just prior to the February USDA Agricultural Outlook Forum.) Assumptions made for the United States will affect the projections for Mexico. For the last several years, a modeling framework capturing market and policy dynamics in both countries produced a projections trajectory for the baseline. For marketing years beyond those reported in the *World Agricultural Supply and Demand Estimates* (WASDE) report (e.g., FY 2011 and beyond for this year), forecasting outcomes in U.S. and Mexican sugar markets were derived from pricing linkages between the two countries. Projections made this year are different because of specific assumptions made for the United States.

U.S. sugar supply and use over the long-term projections period are specified to resemble events and policy choices made in FY 2009. Specifically, U.S. producers do not expand area to keep pace with increases in domestic demand; U.S. policymakers do not increase sugar tariff-rate quotas (TRQs) above minimum access levels (as required by the World Trade Organization (WTO) and Free Trade Agreements (FTA)); and U.S. policymakers aim for an ending- year stocks-to-use ratio equal to FY 2009's 13.5 percent. Mexico is assumed to export sugar to the United States to meet this level. When necessary, Mexico is assumed to import sugar from the world market late in the fiscal year to assure sufficient supplies to meet domestic consumption requirements. As mentioned above, optimum ending-fiscalyear stocks in Mexico have been at about 22 percent of total-year deliveries for human consumption. Formal price linkages, the basis of past projections modeling, are absent.

This article presents two separate sets of results. The first is referred to as version A. For Mexico, it assumes sugar exports are made so that the resulting U.S. ending stocks-to-use ratio is 13.5 percent and that Mexico imports from the world market so that its ending stocks-to-consumption ratio equals 22.0 percent in all projection years. The second set is referred to as Version B. Here, prices are linked through trade (as in previous years' projections). Elements of both approaches are developed below. Also, within each set, differing scenarios are run to show alternative assumptions about the future—specifically, the effect of investments made in the Mexican sugar sector (none beyond maintenance assumed in the main baseline) and lower than forecast HFCS use in Mexico. Both are discussed below.

Table 3. Mexico: Sugar production and supply and sugar and HFCS utilization, by fiscal years 2/11/2010.

Fiscal Year (Oct/Sept)	2003/04	2004/05	2005/06	2006/07	2007/08	2008/09	2009/10 1/
			1,000 metric tons				
Beginning stocks	1,194	1,237	1,965	1,294	1,718	1,975	488
Production	5,330	6,149	5,804	5,633	5,852	5,260	5,100
Imports	327	268	240	474	226	160	720
Supply	6,851	7,654	7,809	7,401	7,796	7,395	6,508
Disappearance							
Human consumption	5,380	5,279	5,326	5,133	5,090	5,065	4,900
Other consumption	220	282	323	390	414	475	400
Miscellaneous							
Total	5,600	5,561	5,649	5,523	5,144	5,540	5,300
Exports	14	128	866	160	677	1,367	490
Total use	5,614	5,689	6,515	5,683	5,821	6,907	5,990
Ending stocks	1,237	1,965	1,294	1,718	1,975	488	518
Stocks-to-human consumption	23.0	37.2	24.3	33.5	38.8	9.6	10.6
Stocks-to-use	22.0	34.6	19.9	30.2	33.9	7.1	8.6
HFCS consumption (dry weight)	135	355	667	698	782	653	900

1/ Forecast.

Sources: USDA, Foreign Agricultural Service, Production, Supply and Distribution database (historical data); *World Agricultural Supply and Demand Estimates.*

Production

Table A-1 gives an overview of the Mexican component of the long-term projections model used for analysis. Sugar production comes from processing sugarcane grown in various regions in Mexico (figure A-1).[2] Regional sugarcane area is a function of lagged regional sugarcane prices, indexed to a wholesale price index. Sugar production depends on available processing capacity.[3] Maintenance of processing capacity depends on regional processor returns (sugar sales less grower payments) exceeding processing costs. Sugar yields are a joint function of trend growth and relative prices of fertilizers and chemicals used in production. Regional sugarcane prices derive from the price of sugar in Mexico and regional productivity measures.

Consumption

The USDA bases its projections of Mexican sweetener consumption on analysis of trends in the consumption of sweetener-containing products (SCPs) and in deliveries of sugar for distribution to households and other users (table A-3). For sweeteners in SCPs the procedure starts by gathering data and projections from Euromonitor International for the following SCPs: carbonated beverages, bakery and cereal products, sugar confectionery, chocolate confectionery, chewing gum, ice cream and other dairy products, and processed foods.

Table A1. ERS long-term sugar and sweeteners projections model:factors affecting Mexican sugar supply and utilization.

Beginning stocks	+ Production	+ Imports	= Net exports	+ Deliveries	+ Ending stocks
-- previous year ending stocks level	Sugar processed from: -- Sugarcane > function of lagged regional sugarcane prices, which are derived from estandar sugar price > production dependent on available processing capacity > processing capacity dependent on processor returns (sugar sales revenue less sugarcane grower payments) exceeding processing costs	-- Imports for IMMEX sugar-containing product exports (exog.) plus: Version A -- Mexican Government sets tariff-rate quota so that end year stocks-to-consumption ratio equals 22 percent Version B -- High-Tier Tariff Imports if price = world price + tariff + marketing costs; otherwise, equal to zero	-- dependent on Mexico exportable sugar surplus: * HFCS use in Mexico; and Version A * Mexico exports sugar to the United States so that ending year U.S. stocks-to-use ratio equals 13.5 percent Version B * linkage of U.S. raw and Mexican estandar sugar prices	-- Deliveries for human consumption = industrial end use demand + direct household demand > Growth in per capita sweetener consumption by end use as function of real GDP growth >Sugar use as a fixed share of industrial end user sweetener demand, unless price of sugar draws near to cost of procuring HFCS (then sugar substitutes for HFCS) -- Other deliveries: sugar to Mexico IMMEX program.	Version A *Mexico import policy assures that ratio of ending stocks to sugar consumption equals 22 percent Version B *Ending stocks calculated as residual: Total supply less use

Source: USDA, ERS, Sugar and Sweeteners Team.

The projections are extended through FY 2021 based on Euromonitor projections (through 2013 and 2014), product growth trends, and assumed growth in real Mexican per capita gross domestic product (GDP). The sweetener content of the products is then estimated based on product coefficients used by the Sugar and Sweeteners Team of the Economic Research Service (ERS) in estimating sugar contained in products imported into the United States.

Macroeconomic and demographic assumptions used in USDA's long-term agricultural projections indicate that between 2011 and 2021, Mexican population is projected to grow 10.6 percent and real per capita GDP 29.8 percent. (Also important for pricing linkages in Version B is an 11.6-percent real appreciation of the Mexican peso relative to the dollar.)

The procedure for estimating and projecting sugar deliveries for nonindustrial uses is based on the estimates and projections of sweeteners in SCPs described immediately above. For the estimation of the historical series, estimated sweeteners in SCPs are subtracted from the sum of PSD estimates of sugar deliveries for human consumption and HFCS deliveries for the period 1997-2009. The series is projected forward based on population and GDP growth.

Table A-5 shows sweetener deliveries, by component and total, for the projections period. Per capita sweetener consumption is not shown to grow through 202 1. Sweetener consumption is mostly a matter of increases in population, with real income gains seeming to matter little. These results are from Euromonitor and ERS analysis and should be interpreted as very conservative. Other growth scenarios could be incorporated for sensitivity analysis (although not pursued in this article).

Table A2. ERS long-term sugar and sweeteners projections model, regional sugar production, Mexico.

Area harvested:
Area harvested (i) = ϕ * (Area adjustment(i)) * (Adjustment for production capacity adjustments(i)) for i = CE (Central), GU (Gulf) , NE (Northeast) , NW (Northwest) , PC (Pacific), SO (South)
Area adjustment(i,t) = f(real sugarcane price(t-1),..., real sugarcane price(t-n))
Adjustment for change in production capacity(i,t) = Min(1 -period lag value, φ/(1 +exp(-α*(processor return(t-1)-processor cost(t-1))/processor cost(t-1)) where up = exogeneous increase in process capacity, and processor return is sales revenue less payments to growers.
Sugar and sugarcrop yields:
Sugarcane yield (i) = α0 + α1 *Trend + α2*sugarcane price(t-1)/fertilizer price(-1) Cane sugar yield (i,j) = β0 + β 1*Trend + β 2*Sugarcane yield (i,j)

Source: USDA, ERS, Sugar and Sweeteners Team.

HFCS substitutes for sugar in beverage and food manufacturing. Most HFCS is used in beverages. The framework assumes that, by FY 2013, the Mexican beverage industry can source 75 percent of its sweetener needs from HFCS. The industry is specified to use less HFCS only if domestic sugar prices approach the cost of importing HFCS from the United States. Factors that might contribute to this scenario would be a runup in corn prices, similar to what happened 2 years ago.

Table A3. ERS long-term sugar and sweeteners projections model, sweetener deliveries for consumption, Mexico.

Procedure:
(1) Gather data and projections from Euromonitor International for consumption of sugar-containing products (SCPs)
• Carbonated beverages
• Bakery and cereal
• Sugar confectionery
• Chocolate confectionery
• Gum
• Dairy products (including ice cream)
• Processed foods
(2) Extend projections through FY 2021.
ln(food category variable, per capita) = $\beta 0 + \beta 1$*shift variable (year index) + $\beta 2$*trend+ $\beta 3$*ln(real per capita GDP)
(3) Estimate sweetener content of products.
(4) Derive estimate for sugar deliveries to nonindustrial end users.
Nonindustrial sweetener deliveries = Total sweetener deliveries for consumption - industrial end user sweetener deliveries
ln(nonindustrial deliveries/industrial end user deliveries) = $\beta 0 + \beta 1$ *ln(per capita GDP)
(5) Derive trade off between sugar and high fructose corn syrup (HFCS) Maximum HFCS penetration:A and sugar trade off:B
(A)

Maximum HFCS shares assumed for analysis, by industry:	
Beverages 1/	75.0%
Food manufacturing:	
Bakery and cereal	18.4%
Confectionery	5.1%
Dairy	40.9%
Processed foods	74.4%

1/ share achieved in 2013; interim values 2010: 48.3%; 2011: 57.2%; 2012: 66.1%; 2013: 75.0%.

(B) Flexible share approach: Max coeff (from A)* Flexibility coeff
Flexibility coeff = [1/(1+EXP(-φ*(Estandar price - Cost Variable)/Cost Variable))] Cost Variable = HFCS production cost (U.S.) + transport + processor return

Source: USDA, ERS, Sugar and Sweeteners Team.

Trade, Ending-Year Sugar Stocks, and Prices

Ending-fiscal-year sugar stocks are the difference between total sugar supplies (beginning stocks, production, and imports) and total use (sugar deliveries and exports). The price of Mexican estandar sugar is an inverse function of the ratio of ending-year stocks to sugar consumption. If HFCS is available in Mexico at a lower price than domestic sugar, there is an incentive to substitute HFCS for that sugar, thereby adding more sugar to ending stocks and leading to lower sugar prices, all else constant.

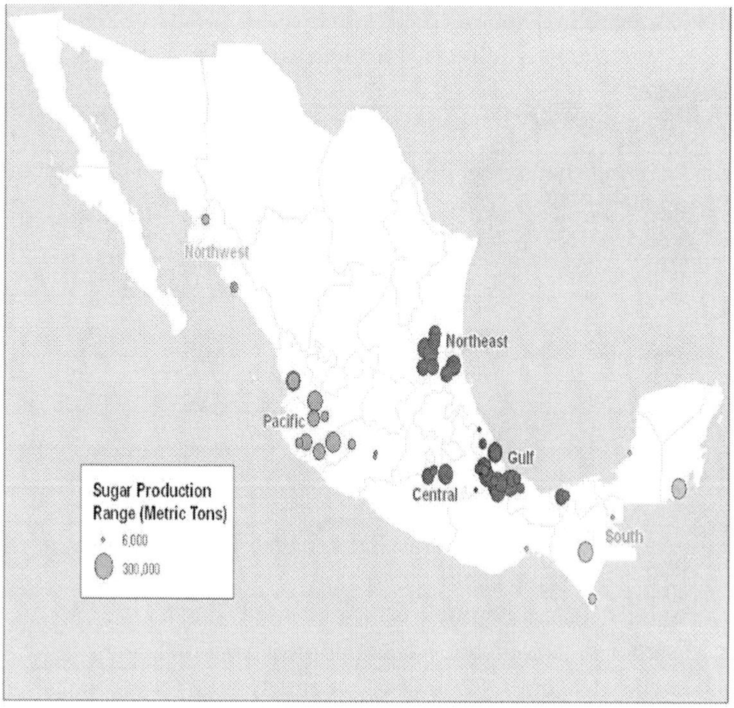

Figure A-1. Sugar in Mexico – Location of Sugarmills by Production Region.

The sweetener-trade phase-in period of the North American Free Trade Agreement (NAFTA) was complete as of January 1, 2008. Trade in all sweeteners between Mexico and the United States is now duty free, with no quantitative restrictions, which implies a link between domestic Mexican sugar prices and those in the United States when Mexico exports sugar to the United States. If sugar prices are higher in the United States than in Mexico, it might be expected that sugar from Mexican stocks would be exported to the United States. A trading equilibrium results when price differences are equal to transport and marketing costs, with adjustments made for the bilateral exchange rate and quality differences.[4]

As explained above, sugar baseline projections this year are based on fixed ending stocks-to-use assumptions (version A), with price parity relationships (version B) used for contrasting alternative scenarios. The framework is shown in figure A-2. Version A assumes Mexico exports to the United States for the U.S. to achieve the 13.5- percent stocks-to-use ratio. U.S. TRQ imports do not rise above WTO and FTA minimum levels. Mexico imports from the third countries to reach its ending stocks target. Prices between the two countries are independent of each other. In Version B, Mexico ships sugar to the United States until price parity (explained above) is achieved. There are no Mexican sugar imports from third countries in version B.

Long-Term Projections of Sugar Supply and Use in Mexico—Version A

Table A-5 shows Mexican sugar supply and use, along with HFCS use and sugar prices, from FY 2011 through 2021. Production averages about 5.8 million MTRV over the projections period. Production increases up to FY 2013 (5.975 million MTRV) because of lagged, earlier-period price effects increasing sugarcane area to 681,000 hectares. After FY 2011, annual estandar prices rise by small amounts, reflecting exchange rate changes with no changes to the underlying stocks-to-consumption ratio.[5] Exports average 1.679 million MTRV, while imports average 729,000 MTRV. With nearly constant prices in both Mexico and the United States, dynamics are largely determined by population growth, especially after FY 2013 when HFCS plateaus at 75 percent of the beverage sweeteners market.

The baseline scenario assumes no net investment increases in the Mexican sugar sector.[6] Incentives for investment should not be ignored. One incentive is the increased stability emanating from access to the U.S. market. Another is also reduced competition in Mexico because of implied marketing allotment limitations for U.S. sugar producers (a perceived restriction to U.S. producers to 85-percent of the U.S. market) and restrictions on policymakers to increase TRQ imports. U.S. producers themselves may see advantages to investment in Mexican sugar facilities, already seen for certain U.S. refiners and traders.

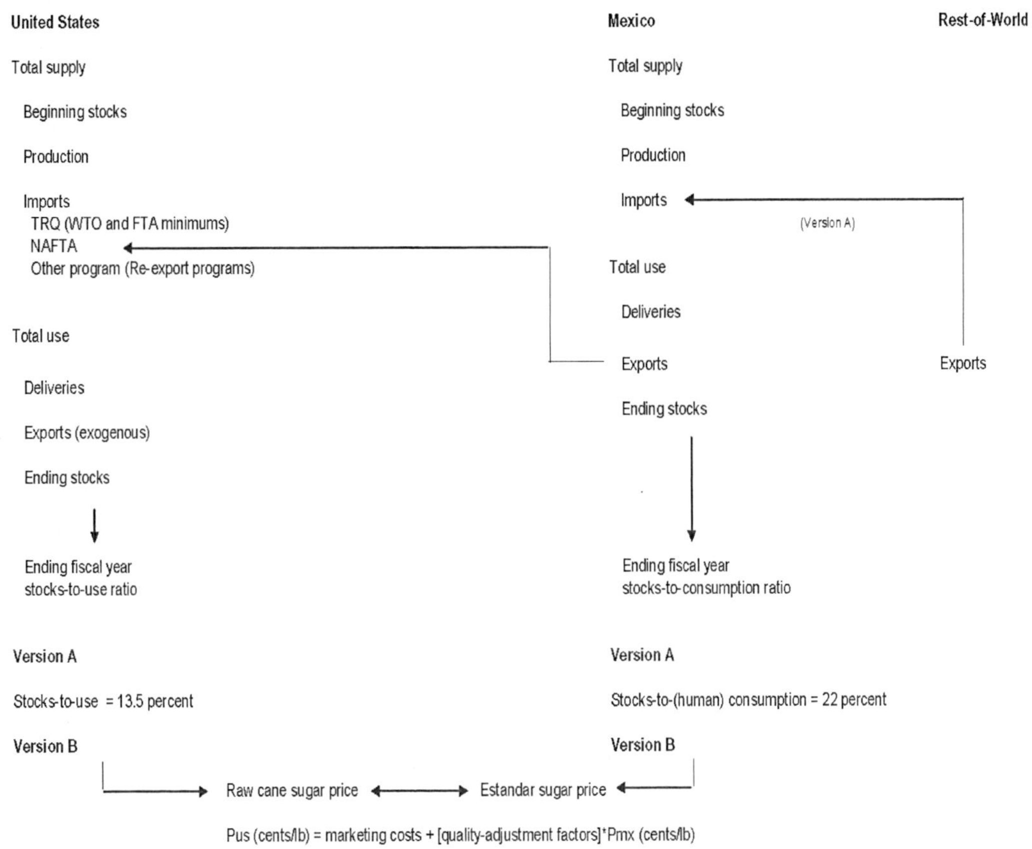

Source: USDA, ERS, Sugar and Sweeteners Group.

Figure A 2. ERS long-term sugar and sweeteners projections model, simplified sugar sector representation.

Table A4. ERS long-term sugar and sweeteners projections model: Mexico, assumptions, sweetener deliveries to end users, marketing year 2011-21.

	2011	2012	2013	2014	2015	2016	2017	2018	2019	2020	2021
	1,000 metric tons, refined value equivalent										
Beverage deliveries	1,639	1,650	1,658	1,668	1,679	1,689	1,699	1,709	1,720	1,730	1,741
Industrial food deliveries	1,123	1,142	1,160	1,178	1,198	1,219	1,240	1,262	1,284	1,307	1,330
Bakery & cereal	754	765	776	787	798	809	820	831	842	853	865
Confectionery	272	276	279	283	288	293	298	303	309	314	320
Dairy	48	49	51	53	55	58	61	64	67	70	74
Processed foods	50	52	54	55	57	59	62	64	66	69	71
Nonindustrial deliveries	2,802	2,828	2,852	2,879	2,906	2,934	2,963	2,991	3,021	3,051	3,082
Total sweetener deliveries	5,564	5,619	5,670	5,726	5,783	5,842	5,902	5,962	6,024	6,088	6,152
Total sweetener deliveries, per capita (kg)	48.9	48.9	48.8	48.7	48.7	48.7	48.7	48.7	48.8	48.8	48.9

Source: USDA, ERS, Sugar and Sweeteners Team.

Table A5. Version A, base sugar supply, utilization, and price projections for Mexico, fiscal years 2011-21.

	2011	2012	2013	2014	2015	2016	2017	2018	2019	2020	2021
	1,000 metric tons, raw value (MTRV) 1/										
Beginning stocks	760	1,056	1,048	1,020	1,028	1,039	1,049	1,060	1,071	1,083	1,094
Area (1,000 hectares)	665	676	681	676	667	655	639	636	625	622	624
Sugar yield (MTRV/hectare)	8.655	8.620	8.776	8.823	8.871	8.919	8.965	9.000	9.064	9.106	9.158
Sugar production	5,753	5,830	5,975	5,966	5,914	5,841	5,726	5,726	5,667	5,661	5,715
Imports	1,073	500	239	323	441	580	763	833	959	1,056	1,249
Supply	7,586	7,385	7,263	7,309	7,383	7,460	7,539	7,619	7,697	7,800	8,057
Disappearance	4,801	4,765	4,637	4,674	4,721	4,769	4,819	4,869	4,921	4,971	5,025
Consumption	4,801	4,765	4,637	4,674	4,721	4,769	4,819	4,869	4,921	4,971	5,025
Other disappearance	0	0	0	0	0	0	0	0	0	0	0
Exports	1,729	1,572	1,606	1,607	1,624	1,642	1,660	1,678	1,693	1,736	1,926
Ending stocks	1,056	1,048	1,020	1,028	1,039	1,049	1,060	1,071	1,083	1,094	1,106
High fructose corn syrup (*dry weight*)	1,038	1,126	1,299	1,319	1,332	1,345	1,358	1,371	1,384	1,401	1,414
Estandar (*cents/lb*)	25.09	23.87	23.99	24.12	24.24	24.36	24.49	24.61	24.74	24.86	24.99

1/ Unless indicated otherwise.

Source: USDA, ERS, Sugar and Sweeteners Team, projections from ERS long-term sugar and sweeteners model.

A second scenario assumes that investment in Mexican producing facilities (except the two factories in the high-cost Northwest region) causes sugarcane area to expand by an annual rate of 2 percent. The version A specification limits the analysis, however. Although production increases by an average of 441,000 MTRV over the projections period (table A-6), imports are displaced exactly by the same amount due to the ending stocks assumption. Nonetheless, investment projections take on more interest under the version B specification.

Alternative Long-Term Projections of Sugar Supply and Use in Mexico—Version B

The version B base is determined through trade flows bringing about sugar pricing parity between the United States and Mexico. The course for both countries' sugar sectors is set by initial supply conditions in the United States and Mexico and by assumptions about U.S. policy. The FY 2010 ending stocks-to-use ratio is a historically low 9.6 percent in the United States, and stocks-to-consumption in Mexico is 15.5 percent. It is assumed that U.S. policymakers do not increase TRQs above minimum access levels and U.S. producers do not expand capacity to increase production potential. No special assumptions are initially made regarding investment in Mexican processing. In these circumstances, annual U.S. stocks-to-use ratios do not rise above the FY 2010 level.

Table A6. Comparison of scenario results for Mexico: Sugar supply, use, and prices, average of fiscal year (FY) 2011-21.

	Scenarios: 1/					
Item	**1 .Base - Version A (V-A)**	**2.Mexico 2% inv. - V-A**	**3.Base - Version B(V-B)**	**4.Mexico 2% inv. - V-B**	**5.HFCS 50% - V-B**	**6.HFCS 50%, 2% inv.- V-B**
	1,000 metric tons, raw value (MTRV) 2/					
Beginning stocks	1,028	1,028	610	1,055	353	667
Area *(1,000 hectares)*	651	701	664	707	687	725
Sugar yield *(MTRV/hectare)*	8.905	8.903	8.941	8.908	9.004	8.950
Sugar production	5,798	6,239	5,937	6,299	6,190	6,492
Imports	729	287	136	136	136	136
Supply	7,554	7,554	6,683	7,490	6,680	7,295
Disappearance	4,816	4,816	4,781	4,817	5,174	5,180
Consumption	4,816	4,816	4,781	4,817	5,174	5,180
Other disappearance	0	0	0	0	0	0
Exports	1,679	1,679	1,338	1,565	1,194	1,411
Ending stocks	1,059	1,059	565	1,108	312	704
Stocks-to-consumption *(ratio)*	0.220	0.220	0.119	0.229	0.061	0.135
High fructose corn syrup *(dry weight)*	1,308	1,308	1,341	1,307	970	964
Estandar *(cents/lb)*	24.49	24.49	27.69	24.54	31.84	27.32

1/ Scenarios:

1. .Base - Version A (V-A): US stocks-to-use = 13.5%, Mx stocks-to-consumption = 22 %

2. Mexico 2% inv., V-A : same as no.1 except investment in sugar industry results in annual 2-percent area increase

3. Base - Version B (V-B): Mexican estandar - U.S. raw sugar price parity

4. Mexico 2% inv., V-B : same as no. 3 except investment in sugar industry results in annual 2-percent area increase

5.Same as no.3 except HFCS 50% of beverage sweetener demand, down from 75 %

6.Same as no. 4 except HFCS 50% of beverage sweetener demand, down from 75 %

2/ Unless indicated otherwise

Source: USDA, ERS, Sugar and Sweeteners Team, projections from ERS long-term sugar and sweeteners model.

Table A7. Version B, base sugar supply, utilization, and price projections for Mexico, fiscal years 2011-21.

Item	2011	2012	2013	2014	2015	2016	2017	2018	2019	2020	2021
	1,000 metric tons, raw value (MTRV) 1/										
Beginning stocks	760	655	588	661	732	756	722	620	518	402	298
Area *(1,000 hectares)*	665	676	681	680	675	667	654	655	648	650	655
Sugar yield *(MTRV/hectare)*	8.655	8.626	8.804	8.860	8.905	8.951	8.997	9.036	9.111	9.168	9.244
Sugar production	5,753	5,834	6,000	6,025	6,009	5,966	5,882	5,918	5,907	5,957	6,054
Imports	136	136	136	136	136	136	136	136	136	136	136
Supply	6,649	6,625	6,723	6,822	6,877	6,858	6,741	6,674	6,561	6,494	6,489
Disappearance	4,801	4,706	4,584	4,635	4,690	4,741	4,788	4,834	4,883	4,935	4,990
Consumption	4,801	4,706	4,584	4,635	4,690	4,741	4,788	4,834	4,883	4,935	4,990
Other disappearance	0	0	0	0	0	0	0	0	0	0	0
Exports	1,193	1,332	1,479	1,454	1,432	1,395	1,332	1,321	1,276	1,261	1,239
Ending stocks	655	588	661	732	756	722	620	518	402	298	259
High fructose corn syrup *(dry weight)*	1,038	1,182	1,348	1,355	1,361	1,372	1,387	1,404	1,420	1,435	1,447
Estandar *(cents/lb)*	25.70	26.47	26.78	26.20	25.93	26.06	26.65	27.74	29.11	30.99	32.94

1/ Unless indicated otherwise.

Source: USDA, ERS, Sugar and Sweeteners Team, projections from ERS long-term sugar and sweeteners model.

Table A-7 shows results for Mexico. Average annual production is only 2.4 percent higher in version B than in version A, although version B FY 2021 production is 5.9 percent higher. Substantially higher prices are not seen in Mexico until FY 2018. Up to then, the version B prices exceed those in version A by less than 2 cents per pound (lb). Although the margins increase after FY 2017, prices affect production only after significant lags. Also, the assumption of no Mexican sugar imports from third countries means less total sugar supply and less sugar to export. Version B's average export level is only about 80 percent of that of version A—1.338 million MTRV per year.

Investments in Mexican processing facilities that imply annual 2-percent area expansion push out total annual supply, which leads to expanded exports—1.565 million MTRV compared with the version B base of 1.338 million MTRV—and increased stocks carryover. Average estandar prices decrease 12.8 percent from the base to 24.54 cents/lb. The resulting stocks-to-consumption ratio at 22.9 percent is close to the optimum level, providing for increased domestic food security.

Reducing HFCS use in Mexico leads to less exportable sugar supply. Two scenarios are run where HFCS constitutes 50 percent of beverage sweetener demand (down from 75 percent in the base) for both the scenario with no investment and the scenario with 2-percent per annum area increase. Exports fall between 144,000 MTRV and 153,000 MTRV as more sugar is used domestically. Estandar prices rise about 28 percent relative to the comparable base scenarios.

Summary

The long-term projections in this article present different perspectives on sugar and sweeteners in Mexico, as well as the United States. Version A represents something of a cooperative fixed-stocks policy in both countries. The ratios used here are arbitrary but do illustrate the approach; other assumptions could be added and modifications made. Version B represents a price-based trade approach. A scenario of restricted HFCS use in Mexico illustrates the effects. Also suggestive are gains available from increased investment in the Mexican sugar sector. Further research regarding costs and tradeoffs with investments in alternative sugar-producing areas could be an extension of this type of analysis.

CONTACTS AND LINKS

Data

Tables from the *Sugar and Sweeteners Yearbook* are available in the Sugar and Sweeteners Briefing Room at http://www.ers.usda.gov/briefing/sugar/. They contain the latest data and historical information on the production, use, prices, imports, and exports of sugar and sweeteners.

Related Websites

Sugar and Sweeteners Outlook http://www.ers.usda.gov/Publications/SSS/
WASDE
http://usda.mannlib.cornell.edu/MannUsda/viewDocumentInfo.do?documented=1194
Sugar Briefing Room, http://www.ers.usda.gov/briefing/Sugar/

E-mail Notification

Readers of ERS outlook reports have two ways they can receive an e-mail notice about release of reports and associated data.

- Receive timely notification (soon after the report is posted on the web) via USDA's Economics, Statistics and Market Information System (which is housed at Cornell University's Mann Library). Go to http://usda.mannlib.cornell.edu/MannUsda/aboutEmailService.do and follow the instructions to receive e-mail notices about ERS, Agricultural Marketing Service, National Agricultural Statistics Service, and World Agricultural Outlook Board products.
- Receive weekly notification (on Friday afternoon) via the ERS website. Go to http://www.ers.usda.gov/Updates/ and follow the instructions to receive notices about ERS outlook reports, Amber Waves magazine, and other reports and data products on specific topics. ERS also offers RSS (really simple syndication) feeds for all ERS products. Go to http://www.ers.usda.gov/rss/ to get started.

Center at (202) 720-2600 (voice and TDD). To file a complaint of discrimination write to USDA, Director, Office of Civil Rights, 1400 Independence Avenue, S.W., Washington, D.C. 20250-9410 or call (800) 795-3272 (voice) or (202) 720-6382 (TDD). USDA is an equal opportunity provider and employer.

End Notes

[1] High fructose corn syrup is measured in dry weight, unless noted otherwise.

[2] These regions are the same as those used by LMC International in its costs of production studies. Those costs of production are used in the ERS modeling of Mexican sugar production.

[3] Table A-2 shows the equations of the production specification.

[4] Estandar sugar produced in Mexico is a very high polarity sugar. Its use in the U.S. refining industry helps that industry reduce its costs.

[5] The modeling framework determines estandar prices for July-September (third quarter of the calendar year). The annual estandar price is projected as a weighted average of the current and previous third-quarter estandar price. The estandar price for FY 2011 is higher than succeeding years because the stocks-to-consumption ratio for FY 2010 is lower than 22 percent, indicating a higher price for the first part of FY 2011.

[6] When the modeling framework was originally developed, more attention was paid to the possibilities of losing processing capacity resulting from trade liberalization scenarios dictating reductions in U.S. sugar loan rates. While the 2008 Farm Act brought increases in loan rates, thus providing for more downward price security, marketing allotments implied limits on the overall size of the U.S. market for U.S. sugar companies.

In: Sugar: Background, Policies and Issues
Editor: Cecilia M. Jordan

ISBN: 978-1-61761-116-2

Chapter 7

SUGAR AND SWEETENERS OUTLOOK (THE USDA PROJECTS U.S. AND MEXICAN SUGAR USE FOR FISCAL YEAR 2010)*

Stephen Haley and Andy Jerardo

The National Agricultural Statistics Service (NASS) forecasts sugar beet acreage intentions for crop year 2009/10 at 1.152 million acres, about 5.7 percent higher than area planted in crop year 2008/09. Assuming normal sucrose levels, normal area harvested-to-planted ratios, and continued improvement in productivity, the U.S. Department of Agriculture (USDA) projects national beet sugar production in crop year 2009/10 at 4.610 million short tons, raw value (STRV). Because 60,000 STRV of this production is expected to occur in September (the last month of fiscal year (FY) 2009), FY 2010 beet sugar production is projected at 4.550 million STRV, according to the *World Agricultural Supply and Demand Estimates* (WASDE) report. This amount is about 6.8 percent higher than the estimated 4.260 million STRV in FY 2009.

The USDA projects FY 2010 cane sugar production at 3.525 million STRV, an increase of 179,000 STRV over that of FY 2009. Because area harvested is not forecast by NASS until the end of June, current cane sugar projections assume the same area harvested for sugar as in the previous year. Florida cane sugar production for FY 2010 is forecast at 1.750 million STRV, assuming the continued dry conditions that have reduced output in the last three growing seasons. The USDA projects FY 2009 Louisiana cane sugar production at 1.400 million STRV. However, hurricane damage in fall of 2008 caused delays in sugarcane planting that could affect the size of the crop. The USDA projects Texas production at 165,000 STRV and Hawaiian production at 210,000 STRV.

Although the raw and refined sugar tariff-rate quotas (TRQ) for FY 2010 have not yet been announced, the USDA projects them at minimum levels implied by existing international commitments to the World Trade Organization (WTO) and at the allocated levels of existing Free Trade Agreements (FTAs). The projection in the May WASDE is,

* This is an edited, reformatted and augmented edition of a United States Department of Agriculture publication, Report SSS-255, dated June 2, 2009.

therefore, at 1.232 million STRV. Projected shortfall at 150,000 STRV is at about the same higher-than-normal level estimated in FY 2009 (155,000 STRV) because several quota-holding countries are diverting sugar to the European Union (EU) before preferential prices available to them are reduced according to the EU's sugar reform schedule.

Until the TRQ is announced, there is no projection for additional specialty sugar. This sugar is mostly organic sugar, and its allocation for FY 2009 was set at 80,000 STRV in addition to the 1,825 STRV included in the minimum access quantity. The USDA projects imports from Mexico at 165,000 STRV.

Other program sugar imports outside the sugar TRQ for FY 2010 are projected to total 400,000 STRV. Other USDA import programs include the Refined Sugar Reexport Program, the Sugar-Containing Products Program, and the Polyhydric Alcohol Program. High-tier tariff sugar imports and sugar in imported syrups are projected at 10,000 STRV. Sugar exports for FY 2009 are forecast at 200,000 STRV. Most of these exports are expected to go to Mexico, where they are used in Mexico's product re-export (IMMEX) program. Almost all such sugar-containing products are expected to be exported to the United States.

Deliveries for domestic food and beverage use for FY 2010 are projected at 10.350 million STRV, a decrease of 150,000 STRV compared with the FY 2009 delivery estimate. Recent increases in Mexican estandar and refinado sugar prices are expected to continue and be sustained through FY 2010. Higher prices are expected to limit Mexican sugar's competitiveness with high fructose corn syrup (HFCS) in the United States.

The projection for Sugar-Containing Product Re-export deliveries is 175,000 STRV. The FY 2010 projection for deliveries is 25,000 STRV for the manufacture of polyhydric alcohol and 35,000 STRV for feed uses, the same as the FY 2009 estimates.

Ending stocks are the difference between supply and use. For FY 2010, ending stocks are projected at 289,000 STRV, implying a seemingly unsustainable low ending-year stocks-to-use ratio of 2.7 percent. For FY 2009, ending stocks are estimated at 1.192 million STRV, implying an ending stocks-to-use ratio of 11.0 percent, which would be lowest level since FY 1975.

The USDA projects 2009/10 Mexican sugar production at 5.500 million metric tons, raw value (MTRV), 50,000 MTRV less than the 5.550 million MTRV of 2008/09. Low 2008/09 sugarcane yields, which are attributed to the high cost of fertilizers, are not expected to improve much in the new year because peso depreciation will keep the costs at high levels. Although Mexican per capita sweetener consumption is expected to grow in 2009/10 due to better economic growth prospects, sugar consumption is forecast down 90,000 MTRV to 5.250 million MTRV because of increased HFCS consumption (800,000 MTRV, dry basis compared with 610,000 MTRV, dry basis in 2008/09). Due to low beginning stocks and mediocre production prospects, Mexico is forecast to import over 500,000 MTRV of sugar to replenish ending-year stock levels. Exports to the United States are forecast to drop to 150,000 MTRV in 2009/10 from the 862,000 MTRV estimated for 2008/09.

U.S. SUGAR

On May 12, 2009, the U.S. Department of Agriculture (USDA) released its latest supply and use estimates for fiscal year (FY) 2009 and first projections for FY 2010 in the *World Agricultural Supply and Demand Estimates* (WASDE) report.

Production

For most of the year, the USDA's Interagency Commodity Estimates Committee (ICEC) for sugar does not project sugar production for the out-year crop. For the most part, the USDA accepts the production estimates and projections provided by beet sugar processors and cane sugar millers to the Farm Service Agency (FSA). However, the processors' and millers' forecasts are not available until June of the preceding crop year. Therefore, in the meantime, WASDE reflects ICEC projections for FY 2010 sugar in May and June.

FY 2010 Beet Sugar Production

The National Agricultural Statistics Service (NASS) forecasts sugar beet acreage intentions for crop year 2009/10 at 1.152 million acres, about 5.7 percent higher than area planted in crop year 2008/09. Area in the Far West region (Idaho, California, and Oregon) is forecast at 201,000 acres, an increase of 21.5 percent. Area in the Great Plains (Colorado, Montana, Nebraska, and Wyoming) is forecast at 155,600 acres, an increase of 10.8 percent. Area in the largest U.S. producing region of the Upper Midwest (Minnesota and North Dakota) is forecast to grow only modestly by 1.4 percent to 657,000 acres. Area in Michigan is forecast at about the same level as last year, 138,000 acres (1,000 acres more than 2008/09).[1]

In making its sugar beet production forecast, the sugar ICEC assumes normal sucrose levels and normal area harvested-to-planted ratios. The national sugar beet yield is forecast at 26.4 tons/acre, only slightly below the 2008/09 record yield of 26.7 tons/acre. Area planted to genetically modified organism (GMO) seed varieties is forecast to grow to about 95 percent of total area, up from about 60 percent last year. The jump in yields in most States took place in 2006/07 and was due mainly to use of rhizomania resistant seed varieties and use of Pancho Beta to control for Curly Top. Fuller adaption of GMO seeds reinforces the trend of substantially higher yields into the coming year.

Assuming trend improvements in processing, the USDA projects national beet sugar production in crop year 2009/10 at 4.610 million short tons, raw value (STRV). Because 60,000 STRV of this production is expected to occur in September (the last month of fiscal year (FY) 2009), FY 2010 beet sugar production is projected at 4.550 million STRV. This amount is about 6.8 percent higher than the estimated 4.260 million STRV in FY 2009.

FY 2010 Cane Sugar Production

The USDA projects FY 2010 cane sugar production at 3.525 million STRV, an increase of 179,000 STRV over that of FY 2009. Because area harvested is not forecast by NASS until the end of June, current cane sugar projections assume the same area harvested for sugar as in the previous year. Florida cane sugar production for FY 2010 is forecast at 1.750 million STRV, assuming the continued dry conditions that have reduced output in the last three growing seasons. The USDA projects FY 2009 Louisiana cane sugar production at 1.400 million STRV. However, hurricane damage in fall 2008 caused delays in sugarcane planting that could affect the size of the crop. The USDA projects Texas production at 165,000 STRV and Hawaiian production at 210,000 STRV.

Trade

Although the raw and refined sugar tariff-rate quotas (TRQ) for FY 2010 have not yet been announced, the USDA projects them at minimum levels implied by existing international commitments to the World Trade Organization (WTO) and at the allocated levels of existing Free Trade Agreements (FTAs). The projection in the May WASDE is, therefore, at 1.232 million STRV (table 1). Projected shortfall at 150,000 STRV is at about the same, higher-than-normal level estimated in FY 2009 because several quota-holding countries are diverting sugar to the European Union (EU) before preferential prices available to them are reduced according to the EU's sugar reform schedule.

Until the TRQ is announced, there is no projection for additional specialty sugar. This sugar is mostly organic sugar, and its allocation for FY 2009 was set at 80,000 STRV in addition to the 1,825 STRV included in the minimum access quantity. The USDA projects imports from Mexico at 165,000 STRV.

Other program sugar imports outside the sugar TRQ for FY 2010 are projected to total 400,000 STRV. Other USDA import programs include the Refined Sugar Reexport Program, the Sugar-Containing Products Program, and the Polyhydric Alcohol Program. High-tier tariff sugar imports and sugar in imported syrups are projected at 10,000 STRV. Sugar exports for FY 2010 are forecast at 200,000 STRV. Most of these exports are expected to go to Mexico, where they are used in Mexico's product re-export (IMMEX) program. Almost all such sugar-containing products are expected to be exported to the United States.

Imports for FY 2009 are estimated at 2.731 million STRV (table 2). Through April 2009, raw sugar TRQ entries equaled 730,826 STRV (662,994 MTRV), or about 68 percent of the expected total (table 3). Sugar from Mexico for the year is estimated at 950,000 STRV, of which 847,193 STRV has entered, or 89 percent of the estimated total. (Last year, only 42 percent of total had entered through the end of April.) About 33 percent of Mexican imports arrived in bulk by ocean vessel. All of this sugar was destined for refining. The remainder was imported in containers, railcars, or trucks. Although some of this sugar was destined to beet processors and cane refiners for further processing, most was for direct consumption or for sale by jobbers.

Refined sugar TRQ entries equaled 233,962 STRV through April 2009. Most of these entries occurred in the last quarter of 2008 and were part of an addition to the FY 2008 refined TRQ, which was announced on August 6, 2008, and which allowed entries until December 31, 2008. Entries of specialty (mostly organic) sugar under the refined sugar TRQ are restricted by tranches that open on pre- specified dates during the fiscal year. The latest tranche opened on May 19, 2009, for 17,230 STRV (15,631 MTRV). The last tranche for the same quantity opens on August 24, 2009.

Table 1. USDA estimates of sugar imports, FY 2010.

Item	Metric tons, raw value	Short tons, raw value
Raw sugar TRQ	1,117,195	1,231,497
Less shortfall attributable to Mexico 1/	-7,258	-8,001
Less other shortfall	-125,866	-138,743
Total raw sugar TRQ	984,071	1,084,753
Refined sugar TRQ		
Allocation to Canada	10,300	11,354
Allocation to Mexico	2,954	3,256
Less Mexican shortfall 1/	-2,954	-3,256
Global	7,090	7,815
Specialty Base	1,656	1,825
Total refined sugar TRQ	19,046	20,994
CAFTA/DR TRQ - calendar 2010	112,700	124,229
Singapore, Bahrain, Jordan	42	46
Peru	2,000	2,205
Total estimate TRQ entries	1,117,859	1,232,227
Mexico	149,687	165,000
Re-export program imports	362,878	400,000
Sugar syrups, high-tier	9,072	10,000
Total projected imports	1,639,496	1,807,227

1/ Total entries from Mexico, quota and nonquota.
Source: USDA, FAS.

U.S. Imports and Exports of Sugar-Containing Products

Sugar in imported products in 2008 is estimated at 1.203 million tons, a 4.5-percent decrease from 2007 and 5.2 percent decrease from 2006 (table 4). The decrease is largely attributable to the reduction of flavored sugar imports from Mexico— Harmonized Tariff Schedule (HTS) 1701.91.4800 and HTS 1701.91.5800—where the elimination of all high-tier sugar tariffs in January 2008 have made this product a less attractive alternative to sugar. Changes in other category imports have been largely offsetting since 2006. This phenomenon contrasts with the rapid import growth period that reached its maximum level in 2006.

Likewise, sugar in exported products has been level the last 3 years, in the area of 582,000 tons. The net inflow of sugar in products is calculated at about 615,000 tons, a decrease from last year of about 84,500 tons.

Sugar in imported products in the first half of FY 2009 is estimated at 561,832 tons, 8.6 percent less than the amount in the corresponding period last year (table 5).

Table 2. USDA estimates of sugar imports, FY 2009.

Item	Metric tons, raw value	Short tons, raw value
Raw sugar TRQ	1,117,195	1,231,497
Less shortfall attributable to Mexico 1/	-7,258	-8,001
Less other shortfall	-133,356	-146,999
Total raw sugar TRQ	**976,581**	**1,076,497**
Refined sugar TRQ		
Allocation to Canada	10,300	11,354
Late FY 2008 entry	15,594	17,189
Allocation to Mexico	2,954	3,256
Less Mexican shortfall 1/	-2,954	-3,256
Global	7,090	7,815
Late FY 2008 entry	68,945	75,998
Re-assigned Mexico refined TRQ	68,278	75,264
Specialty Base	1,656	1,825
Additional	72,575	80,000
Total refined sugar TRQ	**244,438**	**269,446**
CAFTA/DR TRQ - calendar 2009	**110,460**	**121,760**
CAFTA/DR TRQ - calendar 2008	865	953
Singapore, Bahrain, Jordan	42	46
Peru	2,000	2,205
Total estimate TRQ entries	1,334,386	1,470,906
Mexico	861,834	950,000
Re-export program imports	272,158	300,000
Sugar syrups, high-tier	9,072	10,000
Total projected imports	2,477,451	2,730,906

1/ Total entries from Mexico, quota and nonquota.
Source: USDA, FAS.

Deliveries and Ending Stocks

Deliveries for domestic food and beverage use for FY 2010 are projected at 10.350 million STRV, a decrease of 150,000 STRV compared with the FY 2009 delivery estimate. Recent increases in Mexican estandar and refinado sugar prices are expected to continue and be sustained through FY 2010. Higher prices are expected to limit Mexican sugar's competitiveness with high fructose corn syrup (HFCS) in the United States.

In spite of the rise in sugar deliveries of 658,000 STRV between FY 2007 (9.913 million STRV) and FY 2008 (10.571 million STRV) and deliveries of 10.500 million STRV in FY 2009, combined sweetener deliveries of sugar, HFCS, and sugar in imported products have been constant. Figure 1 shows these combined deliveries since 1992 through the first quarter of 2009. The data are organized around an April-March marketing year to include the latest quarter of data. Since 1999/2000 (that is, April 1999-March 2000), the deliveries have averaged 19.036 million tons, refined equivalent value. Values for the latest 3 years are as follows: 2006/07, 19.089 million tons; 2007/08, 19.128 million tons; and 2008/09, 19.115 million tons. Figure 2 shows the data in terms relative to their respective values in 2000/01. Since 2006/07, sugar delivery growth has been offset by decreases in HFCS deliveries. Sugar in product imports stopped growing after March 2007.

The projection for Sugar-Containing Product Re-export deliveries is 175,000 STRV. The FY 2010 projection for deliveries is 25,000 STRV for the manufacture of polyhydric alcohol and 35,000 STRV for feed uses, the same as the FY 2009 estimates.

Ending stocks are the difference between supply and use. For FY 2010, ending stocks are projected at 289,000 STRV, implying a seemingly unsustainable low ending-year stocks-to-use ratio of 2.7 percent. For FY 2009, ending stocks are estimated at 1.192 million STRV, implying an ending stocks-to-use ratio of 11.0 percent, which would be the lowest level since FY 1975.

U.S. Sweetener Demand

The Economic Research Service (ERS) makes calendar year estimates of total sweetener deliveries that are available for food and beverage consumption by U.S. consumers. These sweeteners include refined sugar; the corn sweeteners of high fructose corn syrup (HFCS), glucose syrup, and dextrose; honey; and other edible syrups, including maple syrup and maple sugar.

U.S. deliveries of total sweeteners for human food and beverage use for 2008 are estimated at 20.816 million tons (table 6), representing an increase of 1.3 percent compared with deliveries in 2007. Refined sugar deliveries increased 7.7 percent, whereas corn sweetener deliveries fell by 4.3 percent. Within the corn sweetener category, HFCS deliveries fell for the sixth year in a row, down almost 1 million tons, dry basis, since 2002, or 10.8 percent. Honey deliveries increased by 6.4 percent, and other edible syrups increased slightly.

On a per capita basis, U.S. sweetener deliveries for 2008 were equal to 137.0 pounds, up 0.5 pounds from 2007 but down 14.4 pounds from the per capita high of 151.4 pounds set in 1999.

Sugar contained in imported products has been excluded in estimating U.S. per capita sweetener deliveries. Before 1995, sugar contained in imports was offset by sugar contained in U.S. food exports, therefore indicating only a minor positive adjustment to total deliveries. Beginning in the 1995-96 period, imports of sugar- containing products started increasing at a faster rate than U.S. exports of sugar- containing products. The fourth-to-last column of table 6 shows the addition of sweetener supplies due to net imports of these products. The added amount has grown from 116,000 tons in 1996 to a high of 812,000 tons in 2006. The 2008 total, however, decreased relative to 2006 by 207,000 tons. On a per capita basis, the sugar in net imported products added 4.0 pounds to total per capita sweetener availability in 2008, for a total of 141.0 pounds. This additional amount is the lowest per capita addition to the total since 2002.

Table 3. U.S. monthly sugar imports, FY 2008 and 2009.

Item	Entries by month												Entries to date
	Oct.	Nov.	Dec.	Jan.	Feb	Mar	Apr.	May	June	July	Aug.	Sept.	
	Short tons raw value												
FY 2008													
Mexico	19,358	17,835	37,948	28,948	57,207	51,830	81,496	77,643	66,973	55,272	114,342	85,081	693,932
WTO raw sugar TRQ excluding Mexico	131,642	133,613	97,635	24,074	84,287	66,444	142,800	66,526	108,886	53,257	110,114	33,982	1,053,261
WTO refined sugar TRQ excluding Mexico	14,235	25,756	2,075	1,682	16,412	373	63	15,092	64	42	39,314	87,023	202,130
FTA sugar TRQs	1,509	0	602	11,256	37,536	7,112	6,466	2,060	6,358	15,933	482	9,678	98,992
Re-export program imports	23,347	31,241	55,933	82,984	7,104	45,122	62,849	83,911	24,399	62,160	55,473	30,478	565,002
High-duty sugar	733	981	400	376	585	524	324	320	423	1,216	297	603	6,781
Total	190,824	209,426	194,594	149,320	203,131	171,404	293,999	245,552	207,103	187,880	320,021	246,845	2,620,099
FY 2009													
Mexico	99,165	126,066	128,761	61,658	88,197	173,277	170,070						847,193
WTO raw sugar TRQ excluding Mexico	144,956	95,566	66,200	147,353	49,345	81,735	145,670						730,826
WTO refined sugar TRQ excluding Mexico	121,089	54,356	30,957	22,335	4,896	243	86						233,962
FTA sugar TRQs	953	0	0	21,111	3,849	5,004	12,957						43,875
Re-export program imports	36,703	0	28,904	40	9,509	32,227	9,546						116,928
High-duty sugar	1,063	247	139	7,808	2,239	3,007	n/a						14,502
Total	403,929	276,235	254,961	260,304	158,035	295,493	338,329						1,987,286

Source: U.S. Customs and Border Protection. Source for high-duty sugar is U.S. Census Bureau.

Table 4. Estimated sugar in U.S. product imports and exports, 1995-2008.

Year	Sugar confectionery	Cocoa and cocoa Preparations	Cereal and bakers reparations	Bread, pastry, cakes, etc	Miscellaneous edible preparations	Carbonated soft drinks	Total sugar in imported products	Total sugar in exported products	Net sugar inflow in products
					1,000 short tons				
1995	137,760	66,265	6,286	43,705	68,945	26,405	349,365	243,169	106,196
1996	148,383	75,911	8,580	49,882	60,729	32,456	375,940	317,809	58,131
1997	161,894	92,664	14,273	64,812	68,172	39,403	441,218	356,966	84,252
1998	186,572	97,616	19,110	74,726	91,119	39,811	508,954	390,159	118,795
1999	223,421	111,807	20,116	87,875	118,876	48,165	610,261	371,414	238,847
2000	239,914	130,407	19,548	99,740	120,366	58,745	668,719	392,208	276,511
2001	259,975	160,350	18,097	115,917	127,331	64,961	746,630	442,596	304,034
2002	299,003	193,608	19,419	117,838	140,369	70,852	841,090	470,991	370,099
2003	362,786	208,260	25,139	134,500	150,859	83,440	964,985	459,931	505,053
2004	400,819	220,067	25,082	138,898	186,328	97,731	1,068,925	507,950	560,976
2005	456,969	231,322	26,012	143,742	187,838	109,747	1,155,630	539,237	616,393
2006	499,547	275,449	24,732	148,595	193,692	126,714	1,268,728	596,960	671,768
2007	489,598	276,990	25,081	150,538	189,345	128,811	1,260,363	560,835	699,527
2008	423,394	281,628	25,230	155,565	191,995	125,451	1,203,262	588,293	614,968

Source: USDA, ERS, MTED, Sugar and Sweeteners Team.

Data (in the next-to-last column of table 6) estimated by SRI Consulting and published in their Chemical Economics Handbook (CEH) show the sucrose equivalent of available high-intensity sweeteners saccharin, aspartame, ace sulfame K, sucralose, and cyclamate. The supply of these sweeteners has been growing over time, from 2.91 million tons in 1992 to a high of 4.57 million tons in 2007. Consumption in 2008 is estimated at slightly less than in 2007 at 4.55 million tons. A drop in consumption of diet carbonates of about 1.8 percent contributed to a decrease in the use of aspartame and ace sulfame K. Sucralose, although used in diet carbonates, bucked the downward trend. Saccharin use increased according to trend.

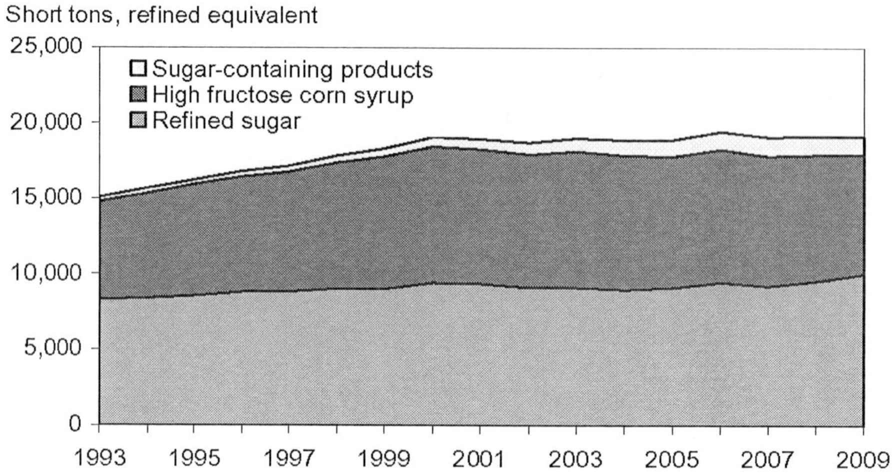

Source: USDA, FSA, SMD (sugar); USDA, ERS, Sugar and Sweetener Team (HFCS, SCP).

Figure 1. Sweetener deliveries, cumulative, April-March marketing year, April 1992-March 2009.

Table 5. Estimated U.S. sugar deliveries and sugar in traded sugar-containing products 1/.

Fiscal year	Oct-Dec	Jan-Mar	Apr-June	July-Sept	FY Total
1,000 short tons, raw value (STRV)					
Domestic sugar deliveries for food and beverage use					
1996	2,379	2,191	2,355	2,519	9,445
1997	2,430	2,143	2,401	2,591	9,565
1998	2,443	2,233	2,428	2,568	9,672
1999	2,458	2,208	2,553	2,655	9,873
2000	2,580	2,318	2,484	2,611	9,993
2001	2,564	2,370	2,486	2,580	10,000
2002	2,474	2,227	2,439	2,645	9,785
2003	2,497	2,183	2,360	2,464	9,504
2004	2,504	2,286	2,368	2,520	9,678
2005	2,547	2,335	2,471	2,666	10,019
2006	2,571	2,436	2,487	2,690	10,184
2007	2,389	2,307	2,535	2,682	9,913
2008	2,513	2,513	2,665	2,879	10,571
2009	2,754	2,437			

Table 5. (Continued)

Fiscal year	Oct-Dec	Jan-Mar	Apr-June	July-Sept	FY Total
Estimated sugar in imported sugar-containing products					
1996	99	85	95	110	389
1997	112	100	119	128	459
1998	125	115	138	151	529
1999	140	140	163	177	620
2000	173	162	177	191	704
2001	185	174	195	216	769
2002	215	192	223	250	879
2003	236	226	256	284	1,002
2004	266	251	288	315	1,119
2005	291	277	298	340	1,205
2006	322	313	358	352	1,345
2007	334	304	321	352	1,311
2008	323	292	324	336	1,274
2009	299	263	0	0	
1996	97	85	90	103	376
1997	103	98	102	108	411
1998	109	91	98	103	401
1999	106	96	99	109	409
2000	116	104	107	128	456
2001	134	115	129	130	508
2002	130	112	118	125	485
2003	138	123	130	140	531
2004	150	137	140	148	575
2005	152	142	160	161	616
2006	175	143	150	150	618
2007	157	145	150	156	608
2008	179	166	173	195	714
2009	193	170	0	0	
Estimated sugar in USDA sugar-containing product re-export program					
1996	21	20	30	32	104
1997	22	68	22	45	157
1998	21	24	32	46	123
1999	44	58	35	32	169
2000	21	21	22	22	86
2001	18	21	29	30	98
2002	40	39	35	42	156
2003	43	44	49	47	183
2004	35	28	40	39	142
2005	28	24	37	33	121
2006	25	25	23	32	106
2007	31	43	55	40	169
2008	35	27	37	41	141
2009	17	30			

Table 5. (Continued)

Fiscal year	Oct-Dec	Jan-Mar	Apr-June	July-Sept	FY Total
Estimated sugar deliveries for domestic consumption (adjusted for trade in sugar-containing products)					
1996	2,402	2,211	2,390	2,558	9,561
1997	2,461	2,213	2,439	2,656	9,770
1998	2,480	2,281	2,500	2,662	9,923
1999	2,536	2,311	2,651	2,755	10,253
2000	2,658	2,396	2,576	2,697	10,328
2001	2,632	2,450	2,580	2,697	10,359
2002	2,599	2,346	2,580	2,811	10,335
2003	2,637	2,330	2,534	2,656	10,158
2004	2,655	2,428	2,555	2,726	10,364
2005	2,714	2,493	2,646	2,877	10,730
2006	2,743	2,630	2,719	2,924	11,016
2007	2,597	2,510	2,761	2,917	10,785
2008	2,693	2,666	2,853	3,061	11,272
2009	2,877	2,561			

Source: USDA, FSA, *Sweetener Market Data*, (deliveries data); USDA, ERS, MTED, Sugar and Sweeteners Team, (sugar in traded products).

1/ includes Puerto Rico.

MEXICAN SUGAR AND HFCS

The USDA estimates 2008/09 Mexican sugar production at 5.550 million metric tons, raw value (MTRV), or 5.239 million metric tons (mt), tel quel (MTTQ). Although the season-long recovery rate is forecast at a record 11.64 percent (tel quel basis), sugarcane yields have been below average and the total sugarcane crop is estimated at only 45.0 million mt, about 5.1 percent less than expected. The disappointing yields are attributed to lower use of fertilizer and other productivity- enhancing inputs due to their high cost relative to the year before. With deterioration in the value of the Mexican peso, these costs are not expected to be lower in the coming season. Therefore, the production forecast for the 2009/10 crop year is 5.5 million MTRV. Sucrose recovery is not expected to be as high as the 2008/09 level.

The USDA expects no growth in 2008/09 per capita sweetener consumption (the sum of sugar and high fructose corn syrup (HFCS)) due to poor macroeconomic conditions. The rate is projected at 50.76 kilograms. The depreciation of the Mexican peso has deterred import of HFCS, as well as corn used in Mexican HFCS production. Total HFCS consumption is projected at 610,000 mt, dry basis, which implies sugar consumption at 5.035 million MTTQ, or 5.340 million MTRV.

The USDA expects economic growth to recover for the 2009/10 marketing year. Per capita consumption is projected to grow at 0.77 percent to 51.15 kilograms. Higher expected Mexican sugar prices will improve the competitiveness of HFCS. The USDA projects HFCS consumption in 2009/10 at 800,000 mt, dry basis. Projected sugar consumption is, therefore, 4.953 million MTTQ, or 5.250 million MTRV.

Deliveries to Mexico's product re-export (IMMEX) program are projected at 440,000 MTRV in 2008/09 and at 400,000 MTRV in 2009/10. Sugar imported from the United States is expected to comprise 115,000 MTRV of IMMEX sugar in 2008/09 and 180,000 MTRV in 2009/10. The remainder comes from Mexican production.

Low Mexican sugar prices in the first half of 2008/09, due to large beginning stocks of 1.615 million MTRV and a weak peso, led to large sugar exports to the United States. The U.S. spot price of refined beet sugar from the *Milling and Baking News*, Midwest, averaged 35.20 cents/pound (lb) the first half of FY 2009 and the Midwest list price of HFCS-42 averaged 31.20 cents/lb, dry basis. In contrast, Mexico City wholesale estandar sugar averaged 18.15 cents/lb and refinado averaged 22.04 cents/lb in the same period. Exports to the United States have totaled 604,000 MTRV through the end of March 2009 and are forecast to total 862,000 MTRV for the entire year. The heavy flows of exports are now contributing to a rise in Mexico City prices in April: 22.33 cents/lb for estandar and 24.30 cents/lb for refinado.

Given expected domestic and export demand, sugar stocks would sink to below 650,000 MTRV, or below 11.9 percent of projected consumption. Because the 2009/10 sugar harvest does not begin until mid-November, these stock levels are considered too low to meet consumer needs in the last quarter of the calendar year.

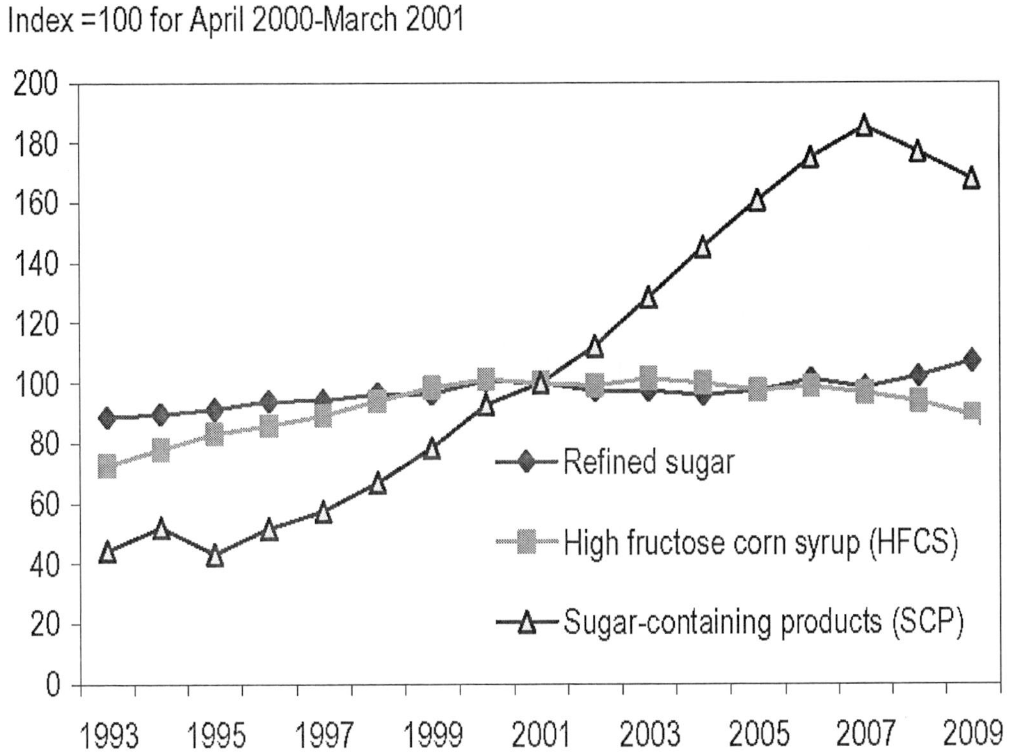

Sources: USDA, FSA, SMD (sugar); USDA, ERS, Sugar and Sweeteners Team (HFCS, SCP).

Figure 2. Sweetener deliveries as proportion of April 2000-March 2001 levels.

Table 6. U.S. caloric sweeteners estimated deliveries for domestic food and beverage use, by calendar year 1/.

Calendar year	U.S. population 2/ (July 1) Millions	Refined sugar	Corn sweeteners				Pure honey	Edible syrups	Total caloric sweeteners	Sugar from net imports	Total caloric sweeteners with sugar-containing products	High intensity sweeteners 3/ (sucrose equivalence)	Total sweeteners
			HFCS	Glucose syrup	Dextrose	Total							
1,000 tons, dry basis													
1992	256.9	8,249	6,652	1,943	461	9,056	126	53	17,483	174	17,657	2,908	20,565
1993	260.3	8,305	7,086	2,050	481	9,617	135	56	18,112	183	18,295	3,032	21,327
1994	263.4	8,478	7,398	2,093	502	9,993	126	54	18,651	153	18,804	3,157	21,960
1995	266.6	8,652	7,676	2,176	528	10,380	120	57	19,209	121	19,331	3,550	22,881
1996	269.7	8,785	7,788	2,216	537	10,541	131	57	19,514	116	19,630	3,695	23,326
1997	272.9	8,861	8,240	2,364	511	11,116	129	58	20,163	197	20,360	3,689	24,049
1998	276.1	8,966	8,552	2,358	502	11,411	130	59	20,566	274	20,840	3,782	24,622
1999	279.3	9,264	8,897	2,281	488	11,666	147	60	21,138	354	21,492	3,877	25,369
2000	282.3	9,253	8,845	2,230	476	11,551	157	61	21,022	304	21,327	3,917	25,243
2001	285.0	9,195	8,920	2,205	469	11,595	134	61	20,986	388	21,373	4,059	25,432
2002	287.7	9,109	9,045	2,224	473	11,741	153	62	21,065	529	21,594	4,193	25,787
2003	290.3	8,859	8,849	2,209	449	11,507	146	63	20,575	621	21,196	4,284	25,479
2004	293.0	9,045	8,779	2,292	487	11,558	130	64	20,797	656	21,453	4,381	25,833
2005	295.7	9,346	8,756	2,261	481	11,497	155	66	21,065	669	21,734	4,414	26,148
2006	298.4	9,323	8,702	2,053	463	11,219	167	66	20,774	812	21,586	4,467	26,053
2007	301.1	9,351	8,479	2,067	448	10,994	141	67	20,553	805	21,358	4,568	25,926
2008 4/	303.8	10,073	8,066	2,037	421	10,524	150	69	20,816	605	21,420	4,552	25,973
Pounds, dry basis													
1992	----	64.2	51.8	15.1	3.6	70.5	1.0	0.4	136.1	1.4	137.5	22.6	160.1
1993	----	63.8	54.5	15.8	3.7	73.9	1.0	0.4	139.2	1.4	140.6	23.3	163.9
1994	----	64.4	56.2	15.9	3.8	75.9	1.0	0.4	141.6	1.2	142.8	24.0	166.7
1995	----	64.9	57.6	16.3	4.0	77.9	0.9	0.4	144.1	0.9	145.0	26.6	171.7
1996	----	65.2	57.8	16.4	4.0	78.2	1.0	0.4	144.7	0.9	145.6	27.4	173.0

Table 6. (Continued)

Calendar year	U.S. population 2/ (July 1) Millions	Refined sugar	Corn sweeteners				Pure honey	Edible syrups	Total caloric sweeteners	Sugar from net imports	Total caloric sweeteners with sugar-containing products	High intensity sweeteners 3/ (sucrose equivalence)	Total sweeteners
			HFCS	Glucose syrup	Dextrose	Total							
1997	----	64.9	60.4	17.3	3.7	81.5	0.9	0.4	147.8	1.4	149.2	27.0	176.2
1998	----	64.9	61.9	17.1	3.6	82.7	0.9	0.4	149.0	2.0	150.9	27.4	178.3
1999	----	66.3	63.7	16.3	3.5	83.5	1.1	0.4	151.4	2.5	153.9	27.8	181.7
2000	----	65.5	62.7	15.8	3.4	81.8	1.1	0.4	148.9	2.2	151.1	27.7	178.8
2001	----	64.5	62.6	15.5	3.3	81.4	0.9	0.4	147.3	2.7	150.0	28.5	178.5
2002	----	63.3	62.9	15.5	3.3	81.6	1.1	0.4	146.5	3.7	150.1	29.2	179.3
2003	----	61.0	61.0	15.2	3.1	79.3	1.0	0.4	141.7	4.3	146.0	29.5	175.5
2004	----	61.7	59.9	15.6	3.3	78.9	0.9	0.4	141.9	4.5	146.4	29.9	176.3
2005	----	63.2	59.2	15.3	3.3	77.8	1.1	0.4	142.5	4.5	147.0	29.8	176.8
2006	----	62.5	58.3	13.8	3.1	75.2	1.1	0.4	139.2	5.4	144.7	29.9	174.6
2007	----	62.1	56.3	13.7	3.0	73.0	0.9	0.4	136.5	5.3	141.8	30.3	172.2
2008 4/	----	66.3	53.1	13.4	2.8	69.3	1.0	0.5	137.0	4.0	141.0	30.0	171.0

1/ Per capita deliveries of sweeteners by U.S. processors and refiners and direct-consumption imports to food manufacturers, retailers, and other end users represent the per capita supply of caloric sweeteners. The data exclude deliveries to manufacturers of alcoholic beverages. Actual human intake of caloric sweeteners is lower because of uneaten food, spoilage, and other losses.

2/ U.S. Census Bureau.

3/ SRI Consulting, *High-Intensity Sweeteners Market Research Report*, Chemical Economics Handbook Program, May 2007.

4/ Preliminary.

Optimal ending-year sugar stocks (September 30) have generally been in the range of 22-2 5 percent of total consumption demand. The USDA assumes Mexico will import sugar to augment fourth-quarter supplies but not quite to the optimal minimum of 1.175 million MTRV (22 percent of 5.340 million MTRV). The USDA projects imports for the buildup of stocks to equal 285,000 MTRV. Total imports for 2008/09 would then total 400,000 MTRV, including the 115,000 MTRV imported from the United States for IMMEX. Ending-year stocks are projected to be 920,000 MTRV, or 17.2 percent of human consumption.

Mexican exports in 2009/10 are expected to be considerably lower than in 2008/09 because low beginning stocks and mediocre production prospects will likely cause Mexican sugar prices to be at higher levels relative to U.S. prices than in 2008/09. The USDA export projection is 150,000 MTRV, about the same level as in FY 2007. The USDA projects that Mexico will import sufficient sugar to assure that ending-year stocks in 2009/10 are equal to 22 percent of projected human consumption. Imports are forecast, therefore, at 535,000 MTRV, and ending-year stocks are projected at 1.155 million MTRV.

HONEY

U.S. honey production climbed close to 161 million pounds in 2008, the first increase since 2004, or 8 percent more than in 2007. Despite the domestic production boost, average honey prices jumped 31 percent as stocks fell and exports rose and as agricultural prices followed the sharp upward movement of commodity prices in 2008 in response to strong world demand. The record-high price of $1.41 per pound exceeded the previous peaks of $1.39 in 2003 and $1.33 in 2002. The 15- percent gain in yield per bee colony more than offset the 6-percent decline in the number of honey-producing bee colonies from more than 2.4 to 2.3 million. Thus, the value of U.S. honey production almost reached $227 million in 2008, sharply up by 42 percent from 2007. As a result, the average farm honey value earned per colony hit a record $98.60, or 51 percent higher than $65.40 in 2007.

South Dakota's 83-percent increase in yield per colony in 2008 boosted the State's production beyond 21.3 million pounds, which was up 61 percent and ranked after North Dakota but surpassed California. As average sale price per pound climbed 31 percent, farm honey value per colony in South Dakota exceeded $127, topping North Dakota's $121.50 and all other major honey-producing States. Indeed, honey bee keepers in the top 10 producing States posted between 9 and 140 percent gains in average farm honey value earned per colony. Although Texas was the only major State in which value of production fell in 2008, farm honey value per colony still gained 9 percent.

The U.S. supply of honey climbed 1 percent to 445 million pounds in 2008, despite the slight decline in imports, which constituted more than 60 percent of total domestic use (fig 3). Thus, per capita use remained at 1.3 pounds, lower than the high of 1.4 pounds in the past two decades. Import volume in 2008 is below the record- high of 278 million pounds in 2006 but equals the average 231 million pounds over the past 5 years. The share of imported honey in domestic consumption is now twice the 30-percent share in 1990/91. The average price of imported honey in 2008 was 95 cents, which is 33 percent lower than the domestic price of $1.41 per pound.

The number one source of imported honey is now Vietnam, surpassing Canada, India, Brazil, China, and Argentina as the major exporters to the United States in recent years. These top six countries supplied 81 percent of U.S. honey imports and 42 percent of total U.S. supply in 2008. Among these suppliers, China's average honey price of 23 cents per pound is 73 percent lower than Vietnam's, the next lowest-priced supplier, and 83 percent below Canada's price, which is the highest at $1.37 per pound. Import tariffs of 1.9 cents per kilogram on honey from China and other countries raise the purchase price to U.S. consumers. In addition, an agricultural fee of 2.2 cents per kilogram is imposed on all imported natural (undiluted, pure) honey. Nevertheless, domestic honey producers face strong competition from imports, the average pre-tax price per pound of which is 46 cents lower than domestic prices.

The 2.301 million bee colonies producing honey in the United States in 2008 were the smallest number in the past two decades. Colony collapse disorder, varroa mites, and other maladies have decimated the domestic bee population. Maintaining colonies at levels sufficient to pollinate local and regional crops has led to importing honey bees and queen bees from Canada, Australia, and New Zealand. This critical resource for crop pollination and concomitant honey production may have to be replenished by imports if the domestic bee colony population does not reverse its downward trend.

Data

Tables from the *Sugar and Sweeteners Yearbook* are available in the Sugar and Sweeteners Briefing Room at http://www.ers.usda.gov/briefing/sugar/. They contain the latest data and historical information on the production, use, prices, imports, and exports of sugar and sweeteners.

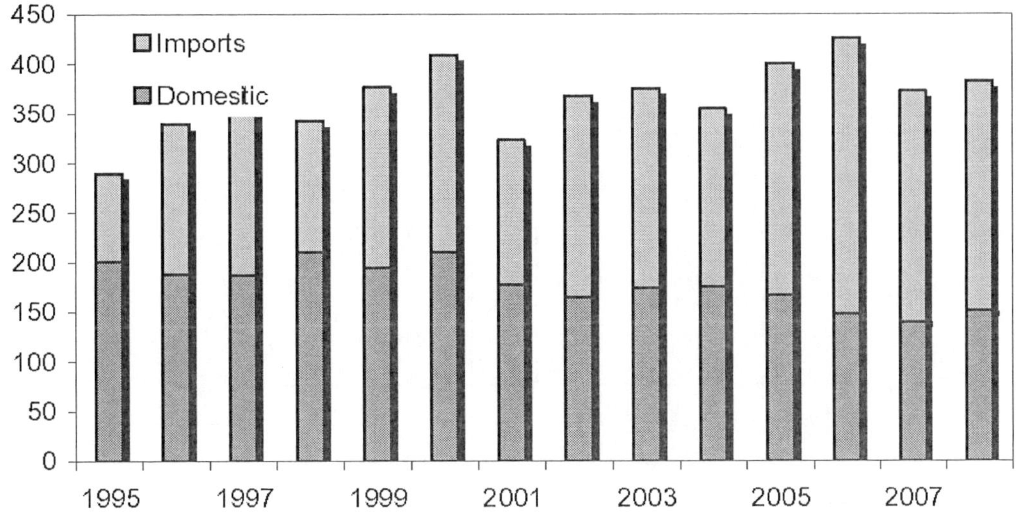

Sources: USDA, NASS, Honey; U.S. Census Bureau.

Figure 3. Share of imports U.S. honey consumption is now 60 percent Million pounds.

Table 7. World refined sugar price, monthly, quarterly, and by calendar and fiscal year 1/.

Year	Jan.	Feb.	Mar.	Apr.	May	June	July	Aug.	Sep.	Oct.	Nov.	Dec.	1st Q.	2nd Q.	3rd Q.	4th Q.	: Calendar	Fiscal
									Cents per pound									
1991	13.39	13.40	13.86	12.90	12.99	13.94	14.73	14.40	13.09	13.03	12.71	12.46	13.55	13.28	14.07	12.73	: 13.41	13.71
1992	12.18	11.92	12.19	12.54	12.89	13.41	13.41	12.96	12.29	11.94	11.68	11.26	12.10	12.95	12.89	11.63	: 12.39	12.67
1993	11.60	11.97	13.05	13.38	13.39	12.64	12.20	13.05	12.90	13.23	13.15	12.97	12.21	13.14	12.72	13.12	: 12.79	12.42
1994	13.14	14.11	15.46	14.92	15.77	16.05	15.54	15.62	15.42	15.46	17.77	18.65	14.24	15.58	15.53	17.29	: 15.66	14.62
1995	18.75	18.17	17.45	16.31	17.05	19.16	20.27	20.01	16.58	17.29	17.64	17.21	18.12	17.51	18.95	17.38	: 17.99	17.97
1996	17.36	17.90	18.14	18.02	17.79	18.00	16.99	16.81	15.74	14.87	14.09	13.95	17.80	17.94	16.51	14.30	: 16.64	17.41
1997	13.87	13.98	14.05	14.19	14.61	14.93	15.07	15.66	14.51	13.58	13.81	13.64	13.97	14.58	15.08	13.68	: 14.33	14.48
1998	13.52	12.78	12.23	11.63	12.00	11.80	11.65	11.62	10.05	10.00	10.78	10.97	12.84	11.81	11.11	10.58	: 11.59	12.36
1999	10.99	10.50	9.85	8.79	9.13	9.93	9.47	9.04	8.28	7.85	7.73	7.61	10.45	9.28	8.93	7.73	: 9.10	9.81
2000	7.70	7.67	7.83	8.66	9.06	10.63	11.38	11.29	11.74	11.76	11.02	10.95	7.73	9.45	11.47	11.24	: 9.97	9.10
2001	11.27	10.65	10.26	10.61	11.71	12.68	12.60	12.08	10.66	10.19	11.27	11.52	10.73	11.67	11.78	10.99	: 11.29	11.35
2002	11.88	10.80	10.81	10.09	10.28	10.02	10.23	10.33	9.68	9.72	10.16	10.25	11.16	10.13	10.08	10.04	: 10.35	10.59
2003	10.64	11.10	10.51	10.14	9.95	9.66	9.84	9.74	8.95	8.39	8.67	9.23	10.75	9.92	9.51	8.76	: 9.74	10.06
2004	9.16	9.54	10.59	11.19	10.78	10.73	11.81	11.80	11.12	11.21	11.27	11.23	9.76	10.90	11.58	11.24	: 10.87	10.25
2005	11.63	12.09	12.02	11.76	11.75	12.61	14.70	14.81	14.60	14.18	13.10	15.00	11.91	12.04	14.70	14.09	: 13.19	12.47
2006	16.92	19.99	20.45	21.35	21.81	20.93	20.95	18.16	17.32	17.92	16.41	15.86	19.12	21.36	18.81	16.73	: 19.01	18.35
2007	15.13	14.92	15.59	14.21	14.94	14.36	14.13	12.87	12.54	12.56	13.00	13.78	15.21	14.50	13.18	13.11	: 14.00	14.91
2008	15.17	16.61	15.79	15.87	14.92	16.35	17.06	17.92	17.52	15.07	15.00	14.27	15.86	15.71	17.50	14.78	15.96	15.55
2009	15.67	17.60	17.83	18.38									17.03					

1/ Contract No. 5, London Daily Price, for refined sugar, f.o.b. Europe, spot, through June 2006. Starting in July 2006, spot price replaced by average of nearest futures month for which an entire month of prices is available. Source: London International Financial Futures and Options Exchange (LIFFE).

Table 8. World raw sugar price, monthly, quarterly, and by calendar and fiscal year 1/.

Year	Jan.	Feb.	Mar.	Apr.	May	June	July	Aug.	Sep.	Oct.	Nov.	Dec.	:	1st Q.	2nd Q.	3rd Q.	4th Q.	:	Calendar	Fiscal
Cents per pound																				
1991	8.88	8.57	9.22	8.55	7.88	9.37	10.26	9.45	9.39	9.10	8.79	9.03	:	8.89	8.60	9.70	8.97	:	9.04	9.26
1992	8.43	8.06	8.22	9.53	9.62	10.52	10.30	9.78	9.28	8.66	8.54	8.15	:	8.24	9.89	9.79	8.45	:	9.09	9.22
1993	8.27	8.61	10.75	11.30	11.87	10.35	9.60	9.30	9.52	10.27	10.10	10.47	:	9.21	11.17	9.47	10.28	:	10.03	9.58
1994	10.29	10.80	11.71	11.10	11.79	12.04	11.73	12.05	12.62	12.75	13.88	14.76	:	10.93	11.64	12.13	13.80	:	12.13	11.25
1995	14.87	14.43	14.58	13.63	13.49	13.99	13.46	13.75	12.72	11.94	11.96	12.40	:	14.63	13.70	13.31	12.10	:	13.44	13.86
1996	12.57	12.97	13.07	12.43	11.94	12.54	12.83	12.33	11.87	11.65	11.29	11.38	:	12.87	12.30	12.34	11.44	:	12.24	12.40
1997	11.13	11.06	11.17	11.50	11.54	12.02	12.13	12.54	12.65	12.86	13.19	12.90	:	11.12	11.69	12.44	12.98	:	12.06	11.67
1998	11.71	11.06	10.66	10.27	10.17	9.33	9.70	9.50	8.21	8.24	8.73	8.59	:	11.14	9.92	9.14	8.52	:	9.68	10.80
1999	8.40	7.05	6.11	5.44	5.83	6.67	6.11	6.39	6.98	6.90	6.54	6.00	:	7.19	5.98	6.49	6.48	:	6.54	7.05
2000	5.64	5.51	5.54	6.48	7.33	8.72	10.18	11.14	10.35	10.96	10.02	10.23	:	5.56	7.51	10.56	10.40	:	8.51	7.53
2001	10.63	10.26	9.64	9.27	9.96	9.80	9.48	8.77	8.60	7.15	7.80	8.02	:	10.18	9.68	8.95	7.66	:	9.12	9.80
2002	7.96	6.81	7.27	7.12	7.33	7.07	8.02	7.86	8.54	8.84	8.87	8.81	:	7.35	7.17	8.14	8.84	:	7.88	7.58
2003	8.56	9.14	8.50	7.92	7.41	6.85	7.18	7.30	6.70	6.74	6.83	6.95	:	8.73	7.39	7.06	6.84	:	7.51	8.01
2004	6.42	7.01	8.23	8.21	8.08	8.41	9.19	8.99	9.10	9.84	9.65	10.19	:	7.22	8.23	9.09	9.89	:	8.61	7.85
2005	10.33	10.51	10.57	10.19	10.23	10.45	10.89	11.09	11.59	12.40	12.86	15.09	:	10.47	10.29	11.19	13.45	:	11.35	10.46
2006	17.27	18.93	18.01	18.21	17.83	16.19	16.61	13.58	12.42	12.09	12.38	12.47	:	18.07	17.41	14.20	12.31	:	15.50	15.78
2007	11.85	11.63	11.44	10.85	10.78	11.05	12.18	11.66	11.61	11.86	11.83	12.47	:	11.64	10.89	11.82	12.05	:	11.60	11.67
2008	13.75	15.16	14.60	13.68	12.23	13.29	14.90	15.58	14.74	12.99	12.87	12.31		14.50	13.07	15.07	12.72		13.84	13.67
2009	13.09	13.90	13.83	14.43										13.61						

1/ Contract No. 11 f.o.b. stowed Caribbean port, including Brazil, bulk spot price, plus freight to Far East.

Source: New York Board of Trade (www.nybot.com).

Table 9. U.S. raw sugar price, duty fee paid, New York, monthly, quarterly, and by calendar and fiscal year 1/.

Year	Jan.	Feb.	Mar.	Apr.	May	June	July	Aug.	Sep.	Oct.	Nov.	Dec.	:	1st Q.	2nd Q.	3rd Q.	4th Q.	:	Calendar	Fiscal
Cents per pound																				
1991	21.86	21.42	21.46	21.23	21.29	21.42	21.25	21.83	22.06	21.76	21.75	21.50	:	21.58	21.31	21.71	21.67	:	21.57	21.89
1992	21.38	21.56	21.36	21.38	21.04	20.92	21.10	21.34	21.55	21.61	21.39	21.11	:	21.43	21.11	21.33	21.37	:	21.31	21.39
1993	20.76	21.16	21.56	21.76	21.36	21.42	21.89	21.85	21.97	21.80	21.87	22.00	:	21.16	21.51	21.90	21.89	:	21.62	21.49
1994	22.00	21.95	21.95	22.08	22.18	22.44	22.72	21.84	21.78	21.58	21.57	22.35	:	21.97	22.23	22.11	21.83	:	22.04	22.05
1995	22.65	22.69	22.46	22.76	23.10	23.09	24.47	23.18	23.21	22.67	22.60	22.63	:	22.60	22.98	23.62	22.63	:	22.96	22.76
1996	22.39	22.68	22.57	22.71	22.62	22.48	21.80	22.51	22.38	22.37	22.12	22.14	:	22.55	22.60	22.23	22.21	:	22.40	22.50
1997	21.88	22.07	21.81	21.79	21.70	21.62	22.04	22.21	22.30	22.27	21.90	21.93	:	21.92	21.70	22.18	22.03	:	21.96	22.00
1998	21.85	21.79	21.74	22.14	22.31	22.42	22.66	22.19	21.92	21.67	21.83	22.19	:	21.79	22.29	22.26	21.90	:	22.06	22.09
1999	22.41	22.38	22.55	22.57	22.65	22.61	22.61	21.24	20.10	19.50	17.45	17.87	:	22.45	22.61	21.32	18.27	:	21.16	22.07
2000	17.70	17.24	18.46	19.43	19.12	19.31	17.64	18.12	18.97	21.15	21.39	20.56	:	17.80	19.29	18.24	21.03	:	19.09	18.40
2001	20.81	21.18	21.40	21.51	21.19	21.04	20.64	21.10	20.87	20.90	21.19	21.43	:	21.13	21.25	20.87	21.17	:	21.11	21.07
2002	21.03	20.69	19.92	19.73	19.52	19.93	20.86	20.91	21.65	21.94	22.22	22.03	:	20.55	19.73	21.14	22.06	:	20.87	20.65
2003	21.62	21.91	22.14	21.87	21.80	21.62	21.32	21.26	21.34	20.92	20.91	20.37	:	21.89	21.76	21.31	20.73	:	21.42	21.76
2004	20.54	20.57	20.86	20.88	20.69	20.03	20.14	20.10	20.47	20.31	20.40	20.55	:	20.66	20.53	20.24	20.42	:	20.46	20.54
2005	20.57	20.36	20.54	21.21	21.96	21.89	21.94	20.49	21.10	21.71	21.83	21.74	:	20.49	21.69	21.18	21.76	:	21.28	20.94
2006	23.61	24.05	23.10	23.56	23.48	23.32	22.44	21.38	21.27	20.22	19.66	19.59	:	23.59	23.45	21.70	19.82	:	22.14	22.62
2007	20.03	20.59	20.85	20.91	21.27	21.33	22.72	21.80	21.42	20.56	20.25	20.12	:	20.49	21.17	21.98	20.31	:	20.99	20.87
2008	20.24	20.21	20.65	20.54	20.83	21.80	23.76	23.15	23.10	21.46	19.83	20.00		20.37	21.06	23.34	20.43		21.30	21.27
2009	20.15	19.83	19.75	21.58										19.91						

1/ Contract No. 14, duty fee paid New York. Average of nearest futures month for which an entire month of prices will be available. For example, April 2001's price average of 21.51 cents is the average of closes for the July 2001 futures during the month of April since there was not a full month of May 2001 futures in April (the May 2001 futures expired April 10, July 2001 became the nearest futures, so July 2001 was used for the entire month of April).

Source: New York Board of Trade (https://www.theice.com/marketdata/reportcenter/reports.htm?reportId=1 0).

Table 10. U.S. wholesale refined beet sugar price, Midwest markets, monthly, quarterly, and by calendar and fiscal year.

Year	Jan.	Feb.	Mar.	Apr.	May	June	July	Aug.	Sep.	Oct.	Nov.	Dec.	:	1st Q.	2nd Q.	3rd Q.	4th Q.	:	Calendar	Fiscal
Cents per pound																				
1991	26.88	26.50	26.50	26.13	26.00	25.75	25.50	25.50	25.00	24.94	24.60	24.50	:	26.63	25.96	25.33	24.68	:	25.65	26.57
1992	25.40	26.50	26.50	26.50	26.40	26.00	25.00	25.00	25.00	24.90	24.13	23.90	:	26.13	26.30	25.00	24.31	:	25.44	25.53
1993	23.25	23.00	23.00	23.50	23.50	23.50	25.50	27.75	27.50	27.50	27.25	26.50	:	23.08	23.50	26.92	27.08	:	25.15	24.45
1994	25.75	25.50	25.50	24.50	24.75	25.25	25.00	25.00	24.70	25.00	25.38	25.50	:	25.58	24.83	24.90	25.29	:	25.15	25.60
1995	25.50	25.50	25.50	25.50	25.13	25.10	24.75	24.75	25.50	25.75	28.13	28.85	:	25.50	25.24	25.00	27.58	:	25.83	25.26
1996	28.69	29.00	29.50	29.50	29.70	29.50	29.50	29.00	29.00	29.00	29.00	29.00	:	29.06	29.57	29.17	29.00	:	29.20	28.84
1997	29.00	29.00	28.13	28.00	28.00	27.50	27.00	26.65	26.38	24.90	25.00	25.50	:	28.71	27.83	26.68	25.13	:	27.09	28.06
1998	25.50	25.50	25.50	25.50	26.00	26.00	26.00	26.00	26.50	26.90	27.00	27.00	:	25.50	25.83	26.17	26.97	:	26.12	25.66
1999	27.20	27.13	27.00	27.00	27.00	27.00	27.00	27.00	27.00	26.00	26.00	25.20	:	27.11	27.00	27.00	25.73	:	26.71	27.02
2000	23.38	22.25	21.50	21.00	19.75	19.00	19.00	19.00	20.70	21.25	21.00	21.80	:	22.38	19.92	19.57	21.35	:	20.80	21.90
2001	23.13	22.75	22.00	20.50	21.38	21.90	22.50	22.50	24.63	25.75	26.20	26.50	:	22.63	21.26	23.21	26.15	:	23.31	22.11
2002	26.75	26.00	25.95	24.63	24.50	24.00	24.00	25.40	26.25	26.75	27.40	27.88	:	26.23	24.38	25.22	27.34	:	25.79	25.49
2003	27.80	26.50	27.13	27.63	28.00	28.00	27.63	25.50	24.00	24.70	23.94	23.63	:	27.14	27.88	25.71	24.09	:	26.21	27.02
2004	23.70	23.50	23.50	23.50	23.50	23.50	23.50	23.50	23.50	23.50	23.38	23.20	:	23.57	23.50	23.50	23.36	:	23.48	23.66
2005	23.50	23.50	23.25	23.80	24.75	25.88	26.00	26.75	40.10	40.00	40.00	36.90	:	23.42	24.81	30.95	38.97	:	29.54	25.63
2006	34.50	36.50	37.10	36.38	35.00	35.00	35.00	34.50	31.20	28.75	27.19	26.10	:	36.03	35.46	33.57	27.35	:	33.10	36.01
2007	25.50	25.00	24.90	25.00	25.00	25.00	25.38	25.60	25.38	25.00	24.50	24.50	:	25.13	25.00	25.45	24.67	:	25.06	25.73
2008	24.13	26.40	28.00	28.00	29.60	33.25	38.00	38.40	38.50	36.20	35.00	35.00		26.18	30.28	38.30	35.40		32.54	29.86
2009	35.00	35.00	35.00	34.25										35.00						

Source: *Milling & Baking News*. Simple average of the lower end of the range of quotations for days in that month. Quotations are weekly.

Table 11. U.S. retail refined sugar price, monthly, quarterly, and by calendar and fiscal year.

Year	Jan	Feb	Mar	Apr	may	June	July	Aug	Sep	Oct	Nov	Dec	1st Q	2nd Q	3rd Q.	4th Q.	Calendar	Fiscal
Cents per pound																		
1991	43.40	43.00	43.40	43.30	43.10	43.20	43.50	42.80	42.20	42.00	41.90	41.80	43.27	43.20	42.83	41.90	42.80	43.08
1992	42.50	42.40	41.90	41.70	41.70	41.50	41.50	41.10	41.00	41.20	41.20	40.60	42.27	41.63	41.20	41.00	41.53	41.75
1993	41.20	41.00	40.60	40.80	40.80	40.30	40.20	40.60	40.40	40.50	40.30	39.80	40.93	40.63	40.40	40.20	40.54	40.74
1994	40.70	40.50	40.10	39.90	40.10	39.70	40.00	39.70	40.30	40.20	39.50	39.20	40.43	39.90	40.00	39.63	39.99	40.13
1995	39.70	39.90	39.80	39.40	39.70	39.50	39.70	39.60	39.80	40.40	40.70	39.80	39.80	39.53	39.70	40.30	39.83	39.67
1996	40.50	40.30	40.60	40.40	41.50	41.80	42.40	42.80	42.60	43.20	42.60	42.80	40.47	41.23	42.60	42.87	41.79	41.15
1997	43.40	42.90	43.10	43.50	43.40	43.60	43.30	43.60	43.60	43.00	42.90	42.80	43.13	43.50	43.50	42.90	43.26	43.25
1998	43.00	42.90	43.30	43.10	42.80	43.10	43.20	43.60	43.20	42.30	42.50	42.70	43.07	43.00	43.33	42.50	42.98	43.08
1999	43.60	43.00	43.70	43.20	43.60	43.10	43.20	43.10	43.70	43.80	42.60	42.60	43.43	43.30	43.33	43.00	43.27	43.14
2000	43.70	43.20	42.90	41.40	42.40	42.80	42.50	42.40	42.40	42.50	41.30	41.40	43.27	42.20	42.43	41.73	42.41	42.73
2001	42.80	43.50	43.70	42.90	43.80	43.50	44.30	43.30	44.20	44.00	42.50	42.50	43.33	43.40	43.93	43.00	43.42	43.10
2002	44.10	43.70	42.60	44.40	42.70	43.00	43.30	43.30	43.70	42.40	41.90	42.10	43.47	43.37	43.43	42.13	43.10	43.32
2003	43.00	42.70	42.70	42.70	43.10	42.90	43.10	43.50	42.60	42.50	41.10	42.20	42.80	42.90	43.07	41.93	42.68	42.73
2004	42.90	42.60	42.60	42.70	42.50	42.50	42.90	42.60	42.60	42.60	42.20	43.00	42.70	42.57	42.70	42.60	42.64	42.48
2005	43.70	43.50	43.30	43.60	42.70	42.80	42.40	43.20	43.70	44.20	44.50	44.90	43.50	43.03	43.10	44.53	43.54	43.06
2006	46.10	46.80	47.10	48.00	49.90	50.40	50.50	51.60	51.50	51.20	51.30	50.60	46.67	49.43	51.20	51.03	49.58	47.96
2007	51.90	51.40	51.80	50.80	51.30	52.10	52.20	51.80	51.80	51.30	51.00	50.30	51.70	51.40	51.93	50.87	51.48	51.52
2008	51.90	51.30	50.40	51.70	52.10	52.50	52.50	53.50	56.30	56.50	52.80	53.40	51.20	52.10	54.10	54.23	52.91	52.07
2009	56.90	56.90	57.10	56.80									56.97					

Source: Bureau of Labor Statistics.

Table 12. U.S. producer price index for corn sweeteners and sugar, monthly.

Year	Jan	Feb	Mar	Apr	may	June	July	Aug	Sep	Oct	Nov	Dec	Annual
Corn sweeteners (liquids and solids), incl. glucose, dextrose, and HFCS, June 1985=100 1/													
2000	98.9	98.0	97.8	98.0	97.9	97.9	97.8	98.0	98.0	97.6	99.2	100.3	98.3
2001	111.3	111.6	111.6	111.5	111.9	111.3	111.3	111.3	112.2	112.3	113.9	114.0	112.0
2002	116.5	120.1	119.7	119.8	117.4	119.6	121.2	121.0	127.4	127.9	125.9	126.5	121.9
2003	130.0	131.4	131.3	131.3	131.5	131.9	--	132.2	131.9	130.6	130.9	130.7	131.3
2004	131.9	132.0	131.9	131.7	131.6	131.7	131.8	131.5	131.6	131.5	131.6	131.6	131.7
2005	133.1	133.3	133.5	133.1	133.1	133.1	133.2	132.9	133.2	137.2	133.1	133.2	133.5
2006	144.5	144.8	145.1	153.4	151.1	151.2	151.2	150.9	150.9	150.9	151.1	151.0	149.7
2007	175.5	176.8	176.8	176.8	176.9	177.1	176.8	176.8	176.5	176.9	177.0	176.6	176.7
Raw cane sugar and other can mill products and byproducts, June 1982=100 1/													
2000	92.7	89.4	95.1	97.4	97.0	99.5	92.7	90.7	95.9	106.1	106.9	103.4	97.2
2001	106.3	107.6	107.6	108.6	107.8	106.1	107.7	107.4	107.1	107.4	108.2	109.8	107.6
2002	109.2	107.0	103.8	103.4	101.4	102.7	106.7	106.9	111.2	111.6	113.9	112.7	107.5
2003	108.8	111.3	113.5	111.6	112.1	111.1	109.8	109.8	108.0	106.8	107.4	105.2	109.6
2004	104.7	104.5	106.4	105.6	105.8	102.7	104.6	103.3	107.1	104.2	104.2	106.5	105.0
2005	106.5	105.6	120.0	121.4	122.9	124.5	125.0	127.2	123.3	125.0	126.4	126.3	121.2
2006	129.5	133.2	129.9	132.9	134.6	135.4	134.2	132.0	132.1	127.5	124.4	123.0	130.7
2007	123.9	125.4	125.9	125.9	127.0	127.2	129.0	127.4	127.6	126.2	124.7	123.0	126.1
2008	124.0	122.3	124.2	124.0	124.5	125.0	129.1	131.0	130.8	128.4	126.8	127.8	126.5
2000	105.4	101.5	100.3	99.1	98.3	98.3	97.7	96.2	95.5	94.7	95.0	94.0	98.0
2001	97.5	97.6	97.8	98.0	99.4	99.5	99.5	100.9	102.0	103.3	105.0	106.8	100.6
2002	108.5	109.8	110.5	111.2	111.1	110.9	111.3	111.3	114.2	114.3	116.1	117.9	112.3
2003	118.7	118.8	119.1	119.5	119.2	119.4	119.3	119.4	113.7	116.6	116.4	116.2	118.0
2004	116.1	116.3	116.4	116.8	116.3	116.6	116.6	116.7	116.9	115.5	115.8	116.1	116.4

Table 12. (Continued)

Year	Jan	Feb	Mar	Apr	may	June	July	Aug	Sep	Oct	Nov	Dec	Annual
2005	116.3	117.8	115.9	116.5	117.3	118.6	118.5	118.4	118.2	122.6	136.0	141.5	121.5
2006	141.9	147.4	148.8	149.0	148.6	149.2	152.0	151.2	146.2	145.0	143.5	138.1	146.7
2007	136.2	136.5	133.8	132.9	129.4	126.6	126.2	126.1	125.9	126.3	124.3	123.9	129.0
2008	121.3	121.5	123.0	124.2	127.6	130.1	131.2	142.4	141.8	140.8	139.2	139.3	131.9
Refined cane sugar and byproducts, June 1982=100 1/													
2000	124.7	121.8	121.7	119.8	120.4	119.8	120.5	119.2	117.5	113.9	113.2	114.4	118.9
2001	112.8	117.5	116.2	114.6	115.1	115.3	115.6	116.6	115.5	115.2	115.2	116.3	115.5
2002	117.4	117.9	121.0	122.3	119.7	121.2	121.3	120.8	120.8	121.0	119.5	120.1	120.2
2003	119.1	122.3	122.8	122.9	122.9	123.5	123.8	124.5	125.5	124.3	122.3	123.4	123.1
2004	120.5	120.4	121.6	121.6	123.0	124.3	123.3	123.5	123.1	123.6	122.5	121.6	122.4
2005	122.8	121.9	121.5	121.4	122.6	123.7	122.4	124.4	125.3	130.4	133.6	140.8	125.9
2006	142.8	146.2	155.5	156.9	155.5	150.7	156.4	153.1	152.3	148.2	143.9	142.3	150.3
2007	144.9	140.4	137.9	136.1	134.9	132.0	132.4	128.5	130.0	124.7	130.1	129.9	133.5

1/ Based on a sample of domestic producers. 2/ Preliminary, all indexes are subject to revision 4 months after original publishing.

Source: Bureau of Labor Statistics.

Table 13. U.S. Consumer Price Index for sugar and selected sweetener-containing products 1/.

Year And month	Sugar and sweets	Sugar and artificial sweeteners	Flour And prepared Flour mixes	Cereals and bakery products	Breakfast cereal	White bread	Cakes, cupcakes, and cookies	Other bakery products
	2/	3/	4/	5/	6/	7/	8/	9/
1982-84=100								
2000	154.0	137.1	160.2	188.3	198.0	199.1	187.9	191.5
2001	155.7	140.3	164.3	193.8	199.7	208.3	192.0	199.1
2002	159.0	143.2	171.0	198.0	203.0	213.4	196.7	203.0

Table 13. (Continued)

Year and month	Sugar and sweets	Sugar and artificial sweeteners	Flour And prepared Flour mixes	Cereals and bakery products	Breakfast cereal	White bread	Cakes, cupcakes, and cookies	Other bakery products
2003	162.0	145.7	178.4	202.8	204.3	218.6	202.8	207.3
2004	163.2	146.9	177.8	206.0	203.5	223.8	206.4	211.8
2005	165.2	149.1	179.6	209.0	203.6	232.1	209.8	211.4
2006	171.5	163.9	182.2	212.8	199.9	238.0	214.2	215.5
2007	176.8	167.1	191.6	222.1	205.0	258.0	221.7	220.5
2008	186.6	170.5	225.8	244.9	212.1	294.2	239.9	236.5
2008								
Jan.	180.2	167.0	202.3	228.7	203.1	273.1	227.9	221.7
Feb.	180.6	167.7	208.8	233.4	205.9	278.9	229.2	227.2
Mar.	182.2	165.4	215.5	236.3	211.4	287.9	232.7	225.0
Apr.	184.9	168.5	224.3	240.0	208.6	291.4	234.5	233.8
May	185.1	169.1	231.1	244.2	211.4	294.6	237.9	237.7
June	185.6	170.0	233.4	245.8	210.6	296.7	240.4	239.1
July	187.1	170.4	236.9	250.3	214.4	302.4	243.4	241.9
Aug.	187.8	172.7	236.1	250.1	213.4	299.5	243.1	243.6
Sep.	189.9	175.7	232.2	250.9	214.9	298.3	244.4	243.8
Oct.	190.5	174.6	231.0	252.8	216.1	301.2	246.9	245.2
Nov.	191.8	171.8	228.4	252.7	218.1	302.1	249.7	238.5
Dec.	193.3	173.0	229.9	253.1	217.9	304.7	248.7	240.9
2009								
Jan.	197.4	178.2	237.4	254.4	217.0	301.1	249.8	247.2
Feb.	196.7	178.5	237.9	254.2	214.8	302.4	249.3	248.6
Mar.	197.1	178.9	233.9	253.7	215.0	304.4	249.6	245.9
Apr.	197.3	175.3	231.2	252.7	217.6	301.8	250.5	244.4

Table 13. (Continued)

Year and month	Nonalcoholic beverages	Carbonated drinks	Noncarbonated juices and drinks	Canned fruits	Candy and chewing gum	Ice cream and related products	Food
	10/	11/	12/	13/	14/	15/	16/
1982-84= 100							
2000	137.8	123.4	104.2	106.9	103.8	164.4	167.8
2001	139.2	125.4	106.0	109.0	104.3	173.4	173.1
2002	139.2	125.6	106.4	111.6	106.2	179.1	176.2
2003	139.8	125.6	106.5	113.7	107.8	175.5	180.0
2004	140.4	127.9	105.7	114.0	108.4	178.3	186.2
2005	144.4	131.9	106.5	118.4	109.5	177.6	190.7
2006	147.4	134.2	109.5	121.5	112.2	179.3	195.2
2007	153.4	140.1	112.9	125.2	116.1	183.4	202.9
2008	160.0	147.0	117.5	135.6	123.2	192.8	214.1
2008							
Jan.	157.9	143.7	116.8	128.3	118.8	189.5	208.6
Feb.	157.8	146.5	116.1	132.3	119.1	190.2	209.2
Mar.	158.1	144.5	116.1	130.2	120.6	188.8	209.4
Apr.	159.7	147.0	117.2	130.7	122.5	190.7	211.1
May	158.3	143.2	117.1	135.2	122.2	190.2	212.1
June	158.3	144.5	115.7	136.2	122.6	190.1	213.2
July	159.3	145.1	117.2	136.3	123.8	187.8	215.3
Aug.	160.1	144.5	117.7	137.3	123.6	192.3	216.4
Sep.	161.5	149.6	118.0	140.6	124.6	194.3	217.7
Oct.	163.7	152.3	119.0	141.0	124.6	199.7	218.7
Nov.	163.0	151.8	119.2	139.8	126.9	200.6	218.7
Dec.	162.8	151.1	120.2	139.1	128.7	199.1	218.8

Table 13 (Continued)

Year and month	Nonalcoholic beverages	Carbonated drinks	Noncarbonated juices and drinks	Canned fruits	Candy and chewing gum	Ice cream and related products	Food
2009 Jan.	164.9	154.9	119.9	141.5	130.3	201.7	219.7
Feb.	164.2	155.5	118.5	142.2	130.2	201.0	219.2
Mar.	165.7	157.9	119.4	144.2	129.8	198.2	218.6
Apr.	162.9	153.8	118.7	140.5	130.7	197.4	218.2

1/ All-urban, unadjusted, U.S. city average.
2/ Series:SEFR, Base: 1982-84=100.
3/ Series: SEFR01, Base: 1982-84=100.
4/ Series: SEFA01, Base: 1982-84=100;
5/ Series: SAF1 11, Base: 1982-84=100.
6/ Series: SEFA02, Base: 1982-84=100.
7/ Series: SS02011, Base: 1982-84=100.
8/ Series: SEFB03, Base: 1982-84=100.
9/ Series: SEFB04, Base: 1982-84=100.
10/ Series: SAF1 14, Base: 1982-84=100.
11/ Series: SEFN01, Base: 1982-84=100.
12/ Series: SEFN03, Base: Dec. 1997=100.
13/ Series: SS13031, Base: Dec. 1997=100.
14/ Series: SEFR02, Base: Dec. 1997=100.
15/ Series: SEFJ03, Base: 1982-84=100.
16/ Series: SAF1, Base: 1982-84=100.
Source: Bureau of Labor Statistics.

Table 14. U.S. Consumer Price Index for sugar and selected sweetener-containing products 1/.

Year	Jan.	Feb.	Mar.	Apr.	May	June	July	Aug.	Sep.	Oct.	Nov.	Dec.:	1st Q.	2nd Q.	3rd Q.	4th Q.:	Fiscal	Calendar
1,000 short tons, raw value																		
U.S. beet sugar for domestic consumption:																		
1994	312	313	370	303	338	406	360	406	437	338	304	282:	995	1,047	1,204	924:	4,254	4,170
1995	301	311	378	311	356	399	384	450	465	404	395	331:	989	1,066	1,300	1,131:	4,279	4,486
1996	316	342	361	343	338	325	350	335	300	333	315	267:	1,018	1,006	984	915:	4,139	3,923
1997	280	272	315	312	326	332	351	373	428	375	316	317:	867	970	1,152	1,009:	3,903	3,997
1998	324	316	362	344	342	401	393	388	409	392	334	308:	1,002	1,087	1,190	1,034:	4,288	4,313
1999	319	325	374	346	361	417	400	427	416	438	392	321:	1,018	1,124	1,244	1,151:	4,419	4,536
2000	320	340	385	341	393	384	348	411	392	412	378	329:	1,045	1,118	1,152	1,119:	4,465	4,433
2001	366	346	401	375	405	403	414	450	408	429	373	311:	1,113	1,183	1,272	1,112:	4,686	4,680
2002	349	315	347	340	375	332	369	365	380	423	396	300:	1,012	1,047	1,114	1,119:	4,285	4,291
2003	315	307	341	338	338	365	380	366	388	395	335	353:	962	1,041	1,134	1,082:	4,255	4,219
2004	359	367	407	387	333	438	408	433	392	423	378	342:	1,133	1,159	1,233	1,143:	4,607	4,668
2005	358	368	395	387	370	416	384	415	449	457	375	337:	1,120	1,173	1,248	1,169:	4,684	4,710
2006	342	306	357	323	362	381	348	406	366	369	329	306:	1,005	1,067	1,120	1,004 :	4,360	4,195
2007	339	330	378	396	414	404	422	456	420	436	402	311	1,047	1,214	1,297	1,149	4,562	4,707
2008	365	401	402	405	422	453	438	424	436	437	352	333	1,167	1,280	1,298	1,122	4,894	4,867
2009	316	282	344										943					
Cane sugar for domestic consumption:																		
1994	332	358	422	361	400	448	411	427	473	443	434	420:	1,112	1,209	1,310	1,298:	4,877	4,929
1995	340	332	432	380	424	438	369	444	423	431	413	381:	1,104	1,243	1,236	1,226:	4,880	4,808
1996	353	376	443	425	452	471	463	488	565	547	500	456:	1,172	1,349	1,515	1,504:	5,262	5,539
1997	397	396	481	444	474	509	462	476	500	525	459	431:	1,274	1,427	1,437	1,416:	5,641	5,553
1998	369	391	470	430	429	481	432	438	506	486	467	451:	1,230	1,339	1,377	1,404 :	5,361	5,349

Table 14. (Continued)

Year	Jan.	Feb.	Mar.	Apr.	May	June	July	Aug.	Sep.	Oct.	Nov.	Dec.:	1st Q.	2nd Q.	3rd Q.	4th Q.:	Fiscal	Calendar
1999	355	379	453	452	500	476	433	490	485	483	481	433:	1,186	1,429	1,407	1,396:	5,427	5,419
2000	383	404	484	425	452	488	455	530	471	534	481	398:	1,272	1,365	1,456	1,414:	5,490	5,508
2001	410	371	470	413	431	458	419	446	417	487	467	384:	1,251	1,302	1,282	1,338:	5,248	5,172
2002	392	378	437	424	458	490	472	486	549	468	444	407:	1,208	1,373	1,507	1,320	5,424	5,407
2003	372	377	467	434	408	475	421	488	415	476	486	413:	1,216	1,317	1,324	1,375:	5,177	5,232
2004	346	393	406	377	415	408	404	448	415	528	466	383:	1,144	1,200	1,268	1,377:	4,987	4,989
2005	377	363	459	400	437	441	418	477	458	476	429	401:	1,199	1,277	1,353	1,306:	5,207	5,136
2006	405	383	440	405	434	466	435	494	441	487	456	384:	1,228	1,305	1,369	1,327:	5,209	5,230
2007	399	363	455	426	426	429	400	497	435	448	470	376	1,217	1,281	1,332	1,295	5,157	5,124
2008	408	411	443	393	443	408	435	432	426	450	434	401	1,262	1,243	1,293	1,285	5,093	5,083
2009	395	392	468										1,255					
Imports to nonreporters																		
1994	5	3	6	1	4	4	5	5	7	10	15	12:	14	9	18	38:	63	78
1995	9	1	1	2	0	0	1	1	4	17	5	0:	12	3	6	22:	59	44
1996	0	0	0	0	0	0	0	1	19	10	1	1:	1	1	20	12:	44	33
1997	1	0	1	2	1	1	1	0	1	15	2	2:	2	4	2	19:	20	27
1998	0	1	0	0	1	1	1	0	0	13	5	1:	1	2	1	19:	23	24
1999	3	1	0	0	0	0	0	0	4	27	3	4:	4	0	4	33:	28	41
2000	0	0	1	0	0	0	0	0	3	26	4	1:	1	0	3	31:	38	36
2001	5	1	0	0	0	0	3	21	3	6	10	8:	6	1	27	24:	65	58
2002	3	1	4	7	1	12	3	6	14	36	19	2:	8	20	24	58:	76	109
2003	3	1	1	1	0	1	1	1	4	25	16	5:	5	2	6	47:	71	60
2004	1	2	6	4	3	3	4	11	4	16	11	1:	9	9	19	27:	84	64
2005	1	1	13	6	4	11	2	6	57	17	24	55:	16	21	65	96:	128	197
2006	92	6	104	26	29	60	71	70	61	32	22	5:	202	115	202	58:	615	577
2007	16	22	4	15	22	4	21	13	19	18	40	11	43	40	53	69	194	206
2008	13	34	37	38	58	46	50	98	140	127	117	104	83	143	288	347	584	862
2009	80	80	80										240					

Table 14. (Continued)

Year	Jan.	Feb.	Mar.	Apr.	May	June	July	Aug.	Sep.	Oct.	Nov.	Dec.:	1st Q.	2nd Q.	3rd Q.	4th Q.:	Fiscal	Calendar
2009	80	80	80										240					
Total sugar for domestic consumption:																		
1994	649	674	798	665	742	857	776	838	918	792	754	714:	2,121	2,265	2,532	2,260:	9,195	9,177
1995	651	644	811	694	780	837	755	894	892	853	813	713:	2,105	2,311	2,542	2,379:	9,218	9,337
1996	670	718	804	769	790	796	813	823	883	891	816	724:	2,191	2,355	2,519	2,430:	9,445	9,496
1997	678	668	797	758	801	841	813	849	928	915	778	750:	2,143	2,401	2,591	2,443:	9,565	9,578
1998	694	707	832	774	772	883	826	826	915	892	806	760:	2,233	2,428	2,568	2,458:	9,672	9,686
1999	676	704	827	798	861	894	833	916	905	947	876	757:	2,208	2,553	2,655	2,580:	9,873	9,996
2000	703	745	870	766	845	872	804	941	867	973	863	728:	2,318	2,484	2,611	2,564 :	9,993	9,977
2001	781	718	871	788	837	861	835	917	828	922	849	703:	2,370	2,486	2,580	2,474 :	10,000	9,911
2002	744	695	788	771	834	834	844	858	943	927	860	709:	2,227	2,439	2,645	2,497 :	9,785	9,808
2003	689	685	809	772	746	841	802	856	807	896	837	771:	2,183	2,360	2,464	2,504 :	9,504	9,511
2004	706	762	819	767	751	850	817	893	810	967	855	726:	2,286	2,368	2,520	2,547:	9,678	9,722
2005	737	732	866	793	811	867	804	897	964	951	828	793:	2,335	2,471	2,666	2,571:	10,019	10,043
2006	839	695	901	755	825	907	853	969	868	888	806	694:	2,436	2,487	2,690	2,389:	10,184	10,002
2007	754	715	838	837	862	837	843	966	873	903	912	698	2,307	2,535	2,682	2,513	9,913	10,037
2008	786	846	881	836	923	907	923	953	1,002	1,013	903	837	2,513	2,665	2,879	2,754	10,571	10,811
2009	790	755	892										2,437					
1,000 short tons, raw value																		
Reexported in products:																		
1994	7	7	7	9	15	15	10	17	17	12	11	5:	20	39	44	28 :	127	131
1995	3	7	7	8	4	7	15	18	5	6	8	7:	18	18	39	21:	103	96
1996	5	5	10	14	8	8	8	13	11	9	7	6:	20	30	32	22:	104	104
1997	32	30	6	6	7	10	12	16	17	7	6	8:	68	22	45	21:	157	156

Table 14. (Continued)

Year	Jan.	Feb.	Mar.	Apr.	May	June	July	Aug.	Sep.	Oct.	Nov.	Dec.:	1st Q.	2nd Q.	3rd Q.	4th Q.:	Fiscal	Calendar
1998	6	9	9	12	10	10	14	15	16	18	15	11:	24	32	46	44:	123	146
1999	26	19	12	14	11	10	15	10	7	9	5	7:	58	35	32	21:	169	145
2000	7	7	7	7	8	7	6	11	5	6	6	7:	21	22	22	18:	86	84
2001	8	5	8	9	10	10	11	11	8	10	16	13:	21	29	30	40 :	98	120
2002	15	13	11	12	12	11	12	14	15	17	12	14:	39	35	42	43:	156	158
2003	16	13	14	14	15	20	19	15	13	16	10	9:	44	49	47	35:	183	175
2004	9	10	9	10	18	11	12	15	13	10	9	9:	28	40	39	28 :	142	135
2005	7	8	9	11	9	17	11	11	11	6	14	6:	24	37	33	25:	121	118
2006	6	10	9	10	6	7	7	10	15	11	8	12:	25	23	32	31:	106	111
2007	18	11	14	17	22	16	16	13	11	8	12	16	43	55	40	35	169	173
2008	11	7	9	14	12	11	17	9	15	8	4	5	27	37	41	17	141	123
2009	9	9	12										30					
Polyhydric alcohol and livestock feed use:																		
1994	1	1	1	1	1	1	1	1	1	1	1	1:	4	3	4	4:	13	14
1995	1	1	2	1	2	2	1	2	2	2	1	1:	4	5	4	4:	17	17
1996	1	1	2	1	2	2	2	2	2	2	1	1:	4	5	5	5:	18	18
1997	1	1	1	2	2	2	2	2	3	2	1	2:	4	6	6	5:	21	21
1998	1	1	2	2	2	1	2	2	2	2	2	2:	4	5	5	6:	20	21
1999	1	2	2	2	2	2	2	2	2	2	2	3	5	6	6	8:	24	26
2000	3	3	3	3	2	2	3	2	3	2	3	2:	9	8	7	7:	32	30
2001	3	3	3	3	4	3	3	4	10	4	3	2:	8	10	17	9:	42	44
2002	3	2	2	2	3	4	4	2	2	2	2	1	7	8	8	5:	33	28
2003	2	2	2	2	2	2	2	2	3	2	3	3:	6	7	7	7:	24	27
2004	3	3	4	4	4	3	4	4	4	4	3	4:	9	11	13	10:	41	44
2005	4	4	4	4	4	5	4	4	5	4	4	5:	12	13	13	13:	48	51

Table 14. (Continued)

Year	Jan.	Feb.	Mar.	Apr.	May	June	July	Aug.	Sep.	Oct.	Nov.	Dec.:	1st Q.	2nd Q.	3rd Q.	4th Q.:	Fiscal	Calendar
2006	5	4	5	4	4	4	4	4	4	5	4	4:	13	12	12	12 :	50	49
2007	4	5	5	4	5	4	4	5	5	5	5	4	14	14	13	14	53	54
2008	6	4	6	5	5	5	5	5	5	6	4	4	16	15	16	14	61	61
2009	4	3	3										11					
Total U.S. sugar deliveries 1/:																		
1994	657	682	806	675	758	873	787	856	936	804	767	720:	2,145	2,307	2,579	2,291:	9,334	9,322
1995	655	653	820	703	786	846	772	914	899	861	823	721:	2,127	2,334	2,585	2,405 :	9,337	9,451
1996	676	724	815	785	800	806	822	838	896	901	824	731:	2,215	2,390	2,557	2,457 :	9,567	9,619
1997	712	699	804	766	810	854	827	867	948	924	785	760:	2,215	2,429	2,641	2,469 :	9,742	9,755
1998	701	718	843	787	784	894	843	843	933	912	823	773:	2,261	2,465	2,619	2,508 :	9,815	9,854
1999	704	725	842	814	875	906	850	928	915	958	883	767:	2,271	2,594	2,693	2,609 :	10,066	10,167
2000	713	755	880	776	855	881	813	954	875	981	871	737:	2,348	2,513	2,641	2,589 :	10,111	10,091
2001	792	726	882	800	851	874	849	932	847	936	869	718:	2,399	2,524	2,628	2,524 :	10,140	10,075
2002	761	710	801	786	848	849	860	874	960	946	874	724:	2,272	2,483	2,694	2,544 :	9,973	9,994
2003	707	701	825	788	764	863	823	873	823	914	849	783:	2,233	2,415	2,519	2,546 :	9,711	9,713
2004	718	775	832	782	773	864	833	912	827	980	866	739:	2,324	2,419	2,572	2,586 :	9,861	9,901

Table 14. (Continued)

Year	Jan.	Feb.	Mar.	Apr.	May	June	July	Aug.	Sep.	Oct.	Nov.	Dec.:	1st Q.	2nd Q.	3rd Q.	4th Q.:	Fiscal	Calendar
2005	748	744	879	808	824	889	820	912	979	960	846	803:	2,370	2,521	2,711	2,609 :	10,188	10,212
2006	850	709	914	768	835	919	865	984	886	903	818	710:	2,474	2,522	2,734	2,432 :	10,339	10,162
2007	776	731	857	858	889	857	862	984	888	916	928	718	2,364	2,604	2,735	2,563	10,134	10,265
2008	803	858	895	855	940	923	945	968	1,023	1,027	912	846	2,556	2,718	2,937	2,785	10,773	10,995
2009	803	767	908										2,478					

Totals may not add due to rounding.

Note: This table commenced in October 1991 when USDA began reporting monthly production data. Puerto Rico data were added beginning October 1993. 1/ Fiscal year totals prior to 1994 differ from supply and use (table) since *WASDE* includes Puerto Rico.

Source: USDA, FSA, *Sweetener Market Data*.

Table 15. U.S. sugar: supply and use, by fiscal year 1/.

Items	1999/00	2000/01	2001/02	2002/03	2003/04	2004/05	2005/06	2006/07	2007/08	2008/09	2009/10 Projection May-09
				1,000 short tons, raw value							
Beginning stocks 2/	1,639	2,216	2,180	1,528	1,670	1,897	1,332	1,698	1,799	1,660	1,192
Total production 3/ 4/	9,050	8,769	7,900	8,426	8,649	7,876	7,399	8,445	8,152	7,606	8,075
Beet sugar	4,974	4,680	3,915	4,462	4,692	4,611	4,444	5,008	4,721	4,260	4,550
Cane sugar	4,076	4,089	3,985	3,964	3,957	3,265	2,955	3,438	3,431	3,346	3,525
Florida	1,966	2,057	1,980	2,129	2,154	1,693	1,367	1,719	1,645	1,569	1,750
Louisiana	1,683	1,585	1,580	1,367	1,377	1,157	1,190	1,320	1,446	1,400	1,400
Texas	105	206	174	191	175	158	175	177	158	152	165
Hawaii	318	241	251	276	251	258	223	222	182	225	210

Table 15. (Continued)

Items	1999/00	2000/01	2001/02	2002/03	2003/04	2004/05	2005/06	2006/07	2007/08	2008/09	2009/10 Projection May-09
Puerto Rico	4	0	0	0	0	0	0	0	0	0	0
Total imports	1,636	1,590	1,535	1,730	1,750	2,100	3,443	2,080	2,620	2,731	1,807
Tariff-rate quota imports 5/	1,124	1,277	1,158	1,210	1,226	1,408	2,588	1,624	1,354	1,471	1,232
Other program Imports	388	238	296	488	464	500	349	390	565	300	400
Nonprogram imports	124	76	81	32	60	192	506	66	701	960	175
Mexico 6/								60	694	950	165
Total supply	12,325	12,575	11,615	11,684	12,070	11,873	12,174	12,223	12,571	11,997	11,074
Total exports 3/	124	141	137	142	288	259	203	422	203	130	200
Quota-exempt for reexport	124	141	137	142	288	259	203	422	203	130	200
Other exports	0	0	0	0	0	0	0				
CCC disposal, for export	0	0	0	0	0	0	0				
Statistical difference 7/	0	0	0	0	0	0	0				
Miscellaneous	-126	123	-24	161	23	94	-67	-132	-66	0	0
CCC disposal, for domestic non-food use	0	10	0	0	0	0	0	0	0	0	0
Refining loss adjustment	0	0	0	0	0	0	0	0	0	0	0
Statistical adjustment 8/	-126	113	-24	161	23	94	-67	-132	-66	0	0
Deliveries for domestic use	10,111	10,132	9,974	9,711	9,862	10,188	10,340	10,135	10,773	10,675	10,585
Transfer to sugar-cont. products for exports under reexport program	86	98	156	183	142	121	106	169	141	115	175
Transfer to polyhydric alcohol, feed	32	33	33	24	41	48	51	53	61	60	60
Deliveries for domestic food and beverage use	9,993	10,000	9,785	9,504	9,678	10,019	10,184	9,913	10,571	10,500	10,350

Table 15. (Continued)

Items	1999/00	2000/01	2001/02	2002/03	2003/04	2004/05	2005/06	2006/07	2007/08	2008/09	2009/10 Projection May-09
Total use	10,090	10,396	10,087	10,014	10,172	10,542	10,476	10,424	10,911	10,805	10,785
Ending stocks 3/	2,216	2,180	1,528	1,670	1,897	1,332	1,698	1,799	1,660	1,192	289
Privately owned	1,919	1,395	1,316								
CCC	297	784	212								
Percent											
Stocks-to-use ratio	22.0	21.0	15.1	16.7	18.7	12.6	16.2	17.3	15.2	11.0	2.7

Note: Numbers may not add due to rounding.

1/ Fiscal year beginning October 1. 2/ Stocks in hands of primary distributors and CCC. 3/ Historical data are from USDA, FSA (formerly ASCS), Sweetener Market Data, and USDA, NASS, Sugar Market Statistics prior to 1992. 4/ Production reflects processors' projections compiled by the FSA.

5/ Actual arrivals under the tariff-rate quota (TRQ) with late entries, early entries, and (TRQ) overfills assigned to the fiscal year in which they actually arrived. The 2009/10 available TRQ assumes shortfall of 150,000 tons.

6/ Does not include Mexico TRQ imports.

7/ Receipts compiled by NASS and FSA Customs data. 8/ Calculated as a residual. Largely consists of invisible stocks change.

Table 16. Net cost of corn starch to U.S. wet-millers, Midwest markets.

	Corn byproducts				Byproduct credits					Net cost	
Period	Yellow dent corn 1/	Corn oil	Corn gluten feed	Corn gluten meal	Corn oil	Corn gluten feed	Corn gluten meal	Total byproduct	Corn	Corn starch	Corn sweetener
	Dollars per bu	*Cents per lb*	*Dollars per short ton*		*----Cents per bushel----*			*Dollars per bu*	*Dollars per bu*	*--Cents per lb--*	
1991	2.40	28.36	101.57	256.07	43.96	68.56	33.93	1.46	0.94	2.97	2.81
1992	2.33	23.89	102.80	259.72	37.03	69.39	34.41	1.41	0.92	2.93	2.77
1993	2.27	21.52	87.99	296.53	33.35	59.39	39.29	1.32	0.95	3.02	2.85
1994	2.40	27.22	89.59	262.50	42.19	60.47	34.78	1.37	1.03	3.26	3.08
1995	2.70	26.67	88.34	244.02	41.33	59.63	32.33	1.33	1.37	4.34	4.10
1996	3.82	24.52	116.25	332.40	38.00	78.47	44.04	1.61	2.22	7.04	6.65
1997	2.67	24.87	83.99	345.22	38.55	56.69	45.74	1.41	1.26	4.00	3.78
1998	2.23	29.90	64.86	260.54	46.34	43.78	34.52	1.25	0.98	3.12	2.95
1999	1.92	23.59	58.77	231.88	36.56	39.67	30.72	1.07	0.85	2.68	2.54
2000	1.88	14.66	51.71	237.63	22.72	34.90	31.49	0.89	0.98	3.13	2.95
2001	1.90	15.75	62.46	253.98	24.41	42.16	33.65	1.00	0.90	2.86	2.70
2002	2.17	20.78	60.33	243.72	32.21	40.72	32.29	1.05	1.12	3.55	3.36
2003	2.29	28.65	72.15	251.36	44.40	48.70	33.31	1.26	1.02	3.25	3.07
2004	2.39	27.59	72.01	308.44	42.76	48.61	40.87	1.32	1.07	3.39	3.20
2005	1.90	28.42	51.33	288.09	44.04	34.65	38.17	1.17	0.73	2.33	2.20
2006	2.41	25.06	59.87	264.89	38.84	40.41	35.10	1.14	1.27	4.02	3.80
2007	3.51	39.23	87.70	402.30	60.81	59.19	53.31	1.73	1.78	5.64	5.33
2008	4.95	63.05	110.97	505.40	97.72	74.90	66.97	2.40	2.55	8.10	7.66
2008											
Jan.	4.55	63.35	135.60	545.00	98.19	91.53	72.21	2.62	1.93	6.13	5.79
Feb.	4.91	74.89	128.75	543.13	116.08	86.91	71.96	2.75	2.16	6.86	6.48
Mar.	5.16	83.55	117.19	561.88	129.50	79.10	74.45	2.83	2.33	7.40	6.99

Table 16. (Continued)

	Corn byproducts				Byproduct credits					Net cost	
Period	Yellow dent corn 1/	Corn oil	Corn gluten feed	Corn gluten meal	Corn oil	Corn gluten feed	Corn gluten meal	Total byproduct	Corn	Corn starch	Corn sweetener
	Dollars per bu	*Cents per lb*	*Dollars per short ton*		*----Cents per bushel----*			*Dollars per bu*	*Dollars per bu*	*--Cents per lb--*	
I	4.87	73.93	127.18	550.00	114.59	85.85	72.88	2.73	2.14	6.79	6.42
Apr.	5.59	87.09	129.10	547.00	134.99	87.14	72.48	2.95	2.64	8.39	7.93
May	5.58	87.29	114.38	529.00	135.30	77.21	70.09	2.83	2.75	8.74	8.26
June	6.55	82.33	112.00	524.38	127.61	75.60	69.48	2.73	3.82	12.14	11.47
II	5.91	85.57	118.49	533.46	132.63	79.98	70.68	2.83	3.07	9.76	9.22
July	5.97	76.64	125.70	554.50	118.79	84.85	73.47	2.77	3.20	10.16	9.60
Aug.	5.04	60.00	108.13	505.00	93.00	72.99	66.91	2.33	2.71	8.61	8.13
Sept.	5.00	48.71	99.30	495.50	75.50	67.03	65.65	2.08	2.92	9.26	8.75
III	5.34	61.78	111.04	518.33	95.76	74.95	68.68	2.39	2.94	9.34	8.83
Oct.	3.90	34.76	91.25	464.13	53.88	61.59	61.50	1.77	2.13	6.76	6.39
Nov.	3.61	31.06	90.63	406.25	48.14	61.18	53.83	1.63	1.98	6.28	5.94
Dec.	3.52	26.88	79.60	389.00	41.66	53.73	51.54	1.47	2.05	6.51	6.15
IV	3.68	30.90	87.16	419.79	47.90	58.83	55.62	1.62	2.05	6.52	6.16
2009											
Jan.	3.81	25.19	96.13	469.38	39.04	64.89	62.19	1.66	2.15	6.82	6.45
Feb.	3.64	29.05	98.88	539.38	45.03	66.74	71.47	1.83	1.81	5.74	5.42
Mar.	3.78	29.64	75.40	424.38	45.94	50.90	56.23	1.53	2.25	7.14	6.75
I	3.74	27.96	90.14	477.71	43.34	60.84	63.30	1.67	2.07	6.57	6.21

Sources: USDA, AMS, http://marketnews.usda.gov/portal/lg; USDA, byproduct credits and net cost calculations.

Note: To calculate the net cost of corn, it is assumed that the average bushel of corn wet-milled in the United States contains 31.5 pounds of recoverable starch, dry weight, as well as 1.55 pounds of corn oil (crude weight), 13.5 pounds of corn gluten feed (commercial weight), and 2.65 pounds of corn gluten meal, (commercial weight). Also, 31.5 pounds of starch, dry weight, produces about 33.33 pounds of corn sweetener (dry weight) because of the chemical gain converting starch to sweetener.

Table 17. U.S. use of field corn, by crop year 1/.

Description	1995/96	1996/97	1997/98	1998/99	1999/2000	2000/01	2001/02	2002/03	2003/04	2004/05	2005/06	2006/07	2007/08	2008/09 2/	2009/10 2/
HFCS	473	492	513	530	540	530	541	532	530	521	529	510	490	470	470
Glucose syrup and dextrose	227	233	229	219	222	218	217	219	228	222	229	239	236	237	240
Total corn sweetener	700	725	742	749	761	748	758	751	758	743	758	749	726	707	710
Corn starch	226	238	246	240	251	247	246	256	272	278	275	272	262	235	250
Wet milling excluding alcohol	926	963	988	989	1,013	995	1,003	1,007	1,030	1,021	1,033	1,021	988	942	960
Alcohol															
Fuel	396	429	481	526	566	628	714	996	1,168	1,323	1,603	2,117	3,026	3,750	4,100
Beverage	125	130	133	127	130	130	131	131	132	133	135	136	135	134	134
Total	521	559	614	653	696	758	845	1,127	1,300	1,456	1,738	2,253	3,162	3,884	4,234
Total	1,447	1,522	1,602	1,642	1,709	1,753	1,848	2,133	2,329	2,477	2,771	3,274	4,149	4,826	5,194
U.S. corn crop	7,374	9,233	9,207	9,759	9,431	9,915	9,503	8,967	10,089	11,807	11,114	10,535	13,038	12,101	12,090
Corn sweetener share	9.49	7.85	8.06	7.67	8.07	7.54	7.97	8.38	7.51	6.29	6.82	7.11	5.57	5.84	5.87
Wet milling excluding alcohol share	12.56	10.43	10.73	10.13	10.74	10.04	10.56	11.23	10.21	8.64	9.30	9.69	7.58	7.78	7.94
Alcohol share	7.07	6.05	6.67	6.69	7.38	7.64	8.89	12.56	12.88	12.33	15.64	21.39	24.25	32.10	35.02
Total	19.62	16.48	17.40	16.83	18.12	17.68	19.45	23.79	23.09	20.97	24.93	31.08	31.82	39.88	42.96

1/ September/August crop year. 2/ Forecast.

Source: USDA, ERS, MTED, Sugar and Sweeteners Team.

Table 18. U.S. high fructose corn syrup (HFCS) deliveries, quarterly, by fiscal and calendar year 1/.

Quarter and year	2000	2001	2002	2003	2004	2005	2006	2007	2008	2009
Quarter										
I	2,129	2,165	2,114	2,122	2,185	2,128	2,195	2,087	2,003	1,946
II	2,482	2,370	2,527	2,469	2,438	2,408	2,431	2,363	2,277	
III	2,400	2,433	2,491	2,408	2,361	2,392	2,345	2,266	2,175	
IV	2,103	2,181	2,161	2,136	2,076	2,130	2,073	2,049	1,907	
Year										
Fiscal	9,200	9,072	9,313	9,160	9,119	9,004	9,102	8,789	8,504	
Calendar	9,114	9,149	9,294	9,135	9,060	9,058	9,045	8,765	8,361	

Table 19. U.S. raw sugar tariff-tate quota (TRQ) World Trade Organizatrion (WTO) allocations and entries by month, fiscal year 2009.

Country	Entries by month												Entries todate	FY 2009 TRQallocation	Entries as share of allocation
	Oct-08	Nov-08	Dec-08	Jan-09	Feb-09	Mar-09	Apr-09	May-09	Jun-09	Jul-09	Aug-09	Sep-09			
	Metric tons, raw value														percent
Argentina	63	448	0	-511	7,462	13,495	0						20,957	45,281	46.3
Australia	31,350	31,350	161	0	110	272	0						63,243	87,402	72.4
Barbados	0	0	0	0	0	0	0						0	7,371	0.0
Belize	0	0	0	0	0	0	0						0	11,583	0.0
Bolivia	0	0	8,424	0	0	0	0						8,424	8,424	100.0
Brazil	0	12,540	22,258	66,904	0	0	16,524						118,226	152,691	77.4
Colombia	2,316	563	540	744	694	142	482						5,481	25,273	21.7
Congo	0	0	0	0	0	0	0						0	7,258	0.0
Costa Rica	0	0	0	0	0	2	3,546						3,548	15,796	22.5
Cote d'Ivoire	0	0	0	0	0	0	0						0	7,258	0.0
Dominican Republic	6,535	166	356	24,174	553	20,919	41,441						94,144	185,335	50.8
Ecuador	0	0	0	0	0	67	0						67	11,583	0.6
El Salvador	0	0	27,098	0	-10,340	10,621	0						27,379	27,379	100.0
Fiji	0	0	0	0	0	154	0						154	9,477	1.6
Gabon	0	0	0	0	0	0	0						0	7,258	0.0
Guatemala	20,824	0	0	13,824	259	5,768	0						40,675	50,546	80.5
Guyana	0	19	19	18	0	0	0						56	12,636	0.4
Haiti	0	0	0	0	0	0	0						0	7,258	0.0
Honduras	10,529	0	0	0	1	0	0						10,530	10,530	100.0
India	0	0	0	0	0	0	0						0	8,424	0.0
Jamaica	0	0	0	0	0	0	0						0	11,583	0.0
Madagascar	0	0	0	0	0	0	0						0	7,258	0.0
Malawi	1,940	1,402	1,358	877	466	251	1,043						7,337	10,530	69.7
Mauritius	0	0	10	12	0	7	0						29	12,636	0.2
Mexico 1/	0	0	0	0	0	0	0						0	7,258	0.0
Mozambique	0	0	63	0	0	0	0						63	13,690	0.5

Table 19. (Continued)

Country	Entries by month												Entries todate	FY 2009 TRQallocation	Entries as share of allocation
	Oct-08	Nov-08	Dec-08	Jan-09	Feb-09	Mar-09	Apr-09	May-09	Jun-09	Jul-09	Aug-09	Sep-09			
Nicaragua	0	0	0	0	0	0	0						0	22,114	0.0
Panama	7,629	14,474	0	0	-23	111	-128						22,063	30,538	72.2
Papua New Guinea	0	0	0	0	0	0	0						0	7,258	0.0
Paraguay	0	0	0	0	0	0	0						0	7,258	0.0
Peru	21,945	0	0	0	0	21,230	0						43,175	43,175	100.0
Philippines	28,372	25,735	-230	27,634	28,735	0	31,242						141,488	142,160	99.5
South Africa	0	0	0	0	0	559	23,661						24,220	24,220	100.0
St. Kitts and Nevis	0	0	0	0	0	0	0						0	7,258	0.0
Swaziland	0	0	0	0	16,849	0	0						16,849	16,849	100.0
Taiwan	0	0	0	0	0	0	0						0	12,636	0.0
Thailand	0	0	0	0	0	403	14,340						14,743	14,743	100.0
Trinidad-Tobago	0	0	0	0	0	0	0						0	7,371	0.0
Uruguay	0	0	0	0	0	0	0						0	7,258	0.0
Zimbabwe	0	0	0	0	0	149	0						149	12,636	1.2
Total	131,503	86,697	60,057	133,676	44,766	74,150	132,151	0	0	0	0	0	663,000	1,117,195	59.3

1/ This amount is also included in Table 60, U.S. Imports of Sugar from Mexico.

Source: U.S. Customs and Border Protection, *Weekly Commodity Status Report*. http://www.fas.usda.gov/smi_arc.asp

Table 20. U.S. imports of sugar and certain sugar-containing products from Mexico, FY 2008 and 2009 1/.

Item	Entries by month												Entries to date
	Oct.	Nov.	Dec	Jan	Feb	Mar	Apr	May	Jun	Jul	Aug	Sept.	
	Metric tons, raw value												
FY 2008													
Sugar for further processing 2/	9,234	0	23,638	16,960	39,220	16,960	40,174	30,210	13,250	0	41,764	8,056	239,466
Sugar not for further processing 2/	8,328	16,180	10,788	9,301	12,677	30,059	33,758	40,227	47,507	50,142	61,965	69,128	390,060
Total sugar	17,562	16,180	34,426	26,261	51,897	47,019	73,932	70,437	60,757	50,142	103,729	77,184	629,526
FY 2009													
Imported in bulk by ocean vessel 2/	16,816	46,158	53,256	0	5,902	71,227	60,208						253,567
Imported in containers, railcars, or trucks	73,145	68,207	63,554	55,935	74,109	85,967	94,077						514,994
Total sugar	89,961	114,365	116,810	55,935	80,011	157,194	154,285						768,561

1/ Beginning 1/1/08, no duty or quota applies to sugar from Mexico. From 10/1/07 - 12/31/07, Mexico had duty-free access of 2,954 metric tons allocated under the refined TRQ and 175,000 metric tons (which included WTO raw sugar allocation to Mexico) established by Presidential Proclamation 8180 issued on September 28, 2007.

2/ Includes imports under Mexico's WTO TRQ allocation for raw sugar. May include entries under U.S. Harmonized Tariff Schedule (HTS) 1701.11.10.00, 1701.11.50.00, 1701.91.10.00, 1701.91.30.00, 1701.99.10.90, and 1701.99.50.90. Raw value is commercial weight multiplied by a factor of 1.06.

3/ Includes entries by ocean and over land.

Source: U.S. Customs and Border Protection. http://www.fas.usda.gov/smi_arc.asp.

Table 21. Bulk sugar prices in Mexico, estandar (standard) sugar.

	Nominal pesos per 50 kg 1/.													
	Jan.	Feb.	Mar.	Apr.	May	June	July	Aug.	Sep.	Oct.	Nov.	Dec.	Calendar	Fiscal
1994	90.85	90.85	90.85	90.85	90.85	90.85	90.85	90.85	90.85	90.85	90.94	91.70	90.93	88.62
1995	91.70	99.33	105.95	106.34	110.92	117.25	117.25	119.80	133.76	140.30	144.91	149.57	119.76	106.32
1996	148.43	152.71	159.88	160.92	162.21	166.86	168.24	171.81	176.29	172.51	160.87	155.08	162.98	158.51
1997	173.20	196.96	187.29	179.11	172.99	179.36	175.96	173.60	176.78	169.63	162.55	162.99	175.87	175.31
1998	178.10	176.01	155.70	163.12	180.02	189.52	186.70	210.43	214.81	215.07	223.54	227.44	193.37	179.13
1999	222.59	214.45	195.14	184.23	184.54	223.55	220.27	207.16	211.56	224.71	242.96	228.98	213.35	210.80
2000	220.61	207.89	207.75	201.33	219.23	216.75	232.14	232.22	230.60	224.57	243.21	263.77	225.01	222.10
2001	248.89	234.25	208.67	189.46	185.45	218.39	222.00	219.07	249.51	249.34	240.23	233.55	224.90	225.60
2002	245.76	244.46	243.44	242.14	240.83	239.15	244.95	248.15	253.40	262.31	266.23	268.39	249.93	243.78
2003	268.50	266.46	265.01	270.04	273.14	278.50	285.05	287.64	294.90	302.40	303.75	319.10	284.54	273.85
2004	309.70	296.25	291.25	298.25	297.25	302.95	317.85	326.20	331.00	329.60	326.05	329.85	313.02	308.00
2005	322.70	312.00	306.00	306.00	305.25	304.10	297.25	300.00	289.00	284.10	283.50	282.50	299.37	310.65
2006	280.40	275.60	273.00	292.50	334.40	353.69	333.00	401.40	440.75	395.85	386.25	374.35	345.10	319.57
2007	361.40	344.95	347.10	341.00	332.30	323.00	321.00	306.50	288.12	280.40	272.12	292.00	317.49	317.96
2008	276.20	260.16	260.97	273.50	255.12	248.87	267.20	261.67	262.60	264.50	264.25	280.60	264.64	267.57
2009	272.75	272.88	289.20	329.63	337.17									
	Real 2000 pesos per 50 kg													
	Jan.	Feb.	Mar.	Apr.	May	June	July	Aug.	Sep.	Oct.	Nov.	Dec.	Calendar	Fiscal
1994	291.09	290.35	289.24	287.68	286.14	284.35	282.58	280.92	279.11	277.91	276.82	276.20	283.53	280.77
1995	260.81	270.52	273.49	251.99	252.37	259.17	254.39	255.71	280.13	287.73	288.90	287.20	268.53	265.79
1996	275.99	277.40	283.88	277.74	275.96	280.77	279.65	282.07	286.42	276.86	253.98	238.66	274.11	281.98
1997	260.65	292.10	275.10	261.28	250.64	257.85	251.08	245.75	247.90	236.06	222.49	220.70	251.80	259.32
1998	235.08	228.17	199.79	208.01	228.51	237.82	232.07	258.32	256.06	251.28	257.56	257.90	237.55	230.26
1999	246.64	235.89	213.78	201.54	201.29	242.12	237.54	222.51	225.78	237.91	255.51	238.94	229.96	232.82
2000	227.62	212.96	211.64	203.45	220.66	216.19	231.47	230.72	228.57	220.84	237.72	256.34	224.85	226.30
2001	239.04	224.96	198.79	180.34	176.92	208.53	212.44	208.82	236.28	235.45	226.80	221.21	214.13	216.75
2002	232.16	230.78	226.27	222.92	219.68	216.46	220.50	222.04	224.84	232.90	234.77	233.63	226.41	224.92

Table 21. (Continued)

	Real 2000 pesos per 50 kg													
	Jan.	Feb.	Mar.	Apr.	May	June	July	Aug.	Sep.	Oct.	Nov.	Dec.	Calendar	Fiscal
2003	230.61	225.60	223.77	230.67	234.29	236.88	241.79	242.88	247.86	251.43	250.47	260.23	239.71	234.64
2004	251.54	236.92	229.38	231.45	229.22	233.43	243.92	248.38	250.64	247.88	244.67	249.04	241.37	243.09
2005	242.80	233.10	227.17	226.06	226.75	226.65	220.25	221.88	212.64	208.61	208.12	206.31	221.69	231.57
2006	202.63	198.62	195.00	205.52	232.00	242.77	228.05	273.25	300.16	270.17	263.92	254.76	238.90	225.09
2007	246.15	233.36	232.92	228.09	223.64	216.49	213.63	206.00	192.13	186.76	180.62	193.81	212.80	231.77
2008	181.78	169.35	168.14	174.93	162.47	157.54	167.62	164.04	163.75	160.80	159.28	168.57	166.52	172.57
2009	163.31	161.88												
	U.S. cents per pound													
	Jan.	Feb.	Mar.	Apr.	May	June	July	Aug.	Sep.	Oct.	Nov.	Dec.	Calendar	Fiscal
1994	26.52	26.40	24.96	24.61	24.85	24.48	24.22	24.37	24.23	24.09	23.96	20.93	24.47	24.66
1995	14.75	15.87	14.18	15.49	16.84	17.07	17.37	17.44	19.21	18.87	17.09	17.65	16.82	18.10
1996	18.00	18.43	19.16	19.54	19.79	20.01	20.04	20.74	21.20	20.23	18.45	17.86	19.45	19.21
1997	20.07	22.90	21.36	20.56	19.86	20.47	20.29	20.24	20.61	19.55	17.83	18.19	20.16	20.24
1998	19.64	18.78	16.49	17.41	19.02	19.27	19.03	20.37	19.07	19.20	20.34	20.83	19.12	18.72
1999	19.94	19.44	18.19	17.72	17.82	21.31	21.33	20.00	20.55	21.29	23.41	22.04	20.25	19.72
2000	21.08	20.01	20.29	19.44	20.92	20.00	22.36	22.72	22.35	21.36	23.21	25.28	21.58	21.32
2001	23.11	21.88	19.72	18.43	18.39	21.80	21.97	21.76	24.02	24.22	23.62	23.14	21.84	21.74
2002	24.33	24.36	24.37	23.97	22.97	22.21	22.72	22.88	22.83	23.58	23.69	23.81	23.48	23.47
2003	22.93	22.09	22.05	23.14	24.17	24.06	24.73	24.20	24.49	24.54	24.72	25.73	23.90	23.58
2004	25.73	24.36	23.98	24.01	23.41	24.12	25.14	25.97	26.14	26.22	26.01	26.72	25.15	24.82
2005	25.99	25.41	24.89	24.98	25.23	25.50	25.27	25.47	24.31	23.79	24.10	24.12	24.92	25.50
2006	24.13	23.85	23.04	24.02	27.35	28.16	27.51	33.49	36.39	32.99	32.11	31.29	28.69	26.66
2007	29.93	28.46	28.33	28.17	27.86	27.05	26.93	25.18	23.69	23.51	22.69	24.42	26.35	28.50
2008	22.98	21.92	22.06	23.60	22.17	21.86	23.73	23.47	22.34	18.95	18.27	18.97	21.69	22.90
2009	17.82	16.95	17.91	22.33										

Source: Servicio Nacional de Informacion de Mercados SNIIM-ECONOMICA.

1/ D.F.- Central de Abasto de Iztapolapa, D.F.

Table 22. Bulk sugar prices in Mexico, refinado (refined) sugar.

	Nominal pesos per 50 kg 1/													
	Jan.	Feb.	Mar.	Apr.	May	June	July	Aug.	Sep.	Oct.	Nov.	Dec.	Calendar	Fiscal
1994	101.83	101.83	101.83	101.83	101.83	101.83	101.83	101.83	101.83	101.83	101.85	102.00	101.85	99.31
1995	102.00	110.46	117.80	118.19	122.85	129.30	129.30	132.15	154.33	154.75	159.84	164.98	132.99	118.50
1996	161.26	167.01	177.07	179.04	178.82	181.29	183.36	186.30	188.39	187.66	186.40	186.42	180.25	173.51
1997	194.96	216.67	216.01	215.62	211.40	211.37	213.08	211.71	210.68	206.09	206.63	204.38	209.88	205.17
1998	209.08	207.25	202.34	198.37	205.43	209.93	212.25	229.75	229.88	244.41	250.01	246.63	220.44	210.12
1999	250.22	251.28	241.93	239.00	233.35	242.83	251.83	243.62	239.71	271.33	267.38	263.02	249.63	244.57
2000	259.02	252.50	250.11	248.45	245.58	237.48	244.47	246.61	245.91	245.09	259.57	271.48	250.52	252.66
2001	276.98	274.56	266.54	256.03	250.26	256.90	260.85	261.87	276.33	279.72	277.48	274.21	267.64	263.04
2002	288.40	283.56	284.03	280.56	278.54	279.34	285.98	292.64	298.51	303.09	306.90	309.50	290.92	283.58
2003	310.81	310.73	308.13	313.20	315.26	320.36	334.24	339.84	363.00	360.00	365.00	360.00	333.38	319.59
2004	352.50	340.00	337.20	340.00	337.50	340.60	345.00	337.40	339.50	339.25	338.20	341.00	340.68	346.23
2005	340.00	339.50	335.60	339.00	338.80	335.75	335.75	333.00	330.75	330.00	335.60	335.10	335.74	337.22
2006	332.80	332.75	350.00	355.00	375.60	412.00	415.25	459.70	532.63	486.20	435.75	424.75	409.37	380.54
2007	412.55	403.50	400.25	398.80	389.94	384.16	383.13	380.84	366.40	351.73	331.99	333.16	378.04	405.52
2008	323.53	313.24	309.41	324.99	316.33	307.83	322.66	329.17	332.43	331.75	330.42	334.94	323.06	324.71
2009	329.50	329.33	339.07	357.34	395.00									
	Real 2000 pesos per 50 kg													
	Jan.	Feb.	Mar.	Apr.	May	June	July	Aug.	Sep.	Oct.	Nov.	Dec.	Calendar	Fiscal
1994	326.27	325.44	324.20	322.45	320.72	318.72	316.73	314.87	312.84	311.50	310.04	307.23	317.58	320.25
1995	290.10	300.83	304.08	280.07	279.52	285.81	280.54	282.07	323.19	317.36	318.65	316.78	298.25	296.25
1996	299.85	303.38	314.40	309.01	304.22	305.05	304.79	305.86	306.08	301.17	294.28	286.89	302.91	308.79
1997	293.39	321.33	317.29	314.54	306.29	303.87	304.05	299.70	295.44	286.79	282.82	276.75	300.19	303.19
1998	275.98	268.67	259.64	252.96	260.76	263.43	263.83	282.04	274.03	285.56	288.06	279.66	271.22	270.64
1999	277.25	276.41	265.04	261.46	254.53	263.00	271.57	261.68	255.83	287.27	281.19	274.47	269.14	270.00
2000	267.25	258.66	254.80	251.06	247.19	236.86	243.76	245.02	243.74	241.02	253.71	263.83	250.57	257.61
2001	266.02	263.67	253.92	243.70	238.75	245.30	249.62	249.61	261.68	264.14	261.97	259.72	254.84	252.57

Table 22. (Continued)

	Real 2000 pesos per 50 kg													
	Jan.	Feb.	Mar.	Apr.	May	June	July	Aug.	Sep.	Oct.	Nov.	Dec.	Calendar	Fiscal
2002	272.44	267.69	263.99	258.29	254.07	252.84	257.43	261.85	264.87	269.10	270.63	269.41	263.55	261.61
2003	266.95	263.09	260.18	267.53	270.42	272.48	283.52	286.95	305.09	299.33	300.98	293.59	280.84	273.78
2004	286.31	271.91	265.57	263.85	260.26	262.44	264.75	256.91	257.08	255.13	253.79	257.46	262.96	273.58
2005	255.81	253.64	249.15	250.44	251.67	250.24	248.78	246.28	243.36	242.31	246.37	244.72	248.56	251.31
2006	240.50	239.80	250.00	249.44	260.58	282.79	284.38	312.93	362.71	331.83	297.75	289.06	283.48	268.04
2007	280.99	272.97	268.59	266.76	262.43	257.48	254.98	255.96	244.33	234.27	220.36	221.13	253.35	273.59
2008	212.93	203.91	199.35	207.86	201.45	194.87	202.41	206.35	207.29	201.68	199.17	201.21	203.21	209.35
2009	197.29	195.37												
	U.S. cents per pound													
	Jan.	Feb.	Mar.	Apr.	May	June	July	Aug.	Sep.	Oct.	Nov.	Dec.	Calendar	Fiscal
1994	29.73	29.59	27.97	27.58	27.85	27.44	27.15	27.32	27.15	27.00	26.84	23.28	27.41	27.64
1995	16.41	17.65	15.77	17.21	18.66	18.82	19.16	19.23	22.16	20.81	18.85	19.47	18.68	20.18
1996	19.56	20.15	21.23	21.75	21.81	21.74	21.84	22.49	22.65	22.01	21.37	21.47	21.51	21.03
1997	22.59	25.19	24.63	24.76	24.26	24.12	24.57	24.68	24.56	23.75	22.66	22.81	24.05	23.69
1998	23.05	22.11	21.42	21.17	21.71	21.35	21.64	22.24	20.41	21.82	22.75	22.58	21.86	22.03
1999	22.41	22.78	22.55	22.99	22.53	23.15	24.38	23.52	23.28	25.71	25.76	25.31	23.70	22.90
2000	24.75	24.30	24.43	23.99	23.44	21.91	23.55	24.13	23.83	23.31	24.77	26.02	24.03	24.26
2001	25.72	25.65	25.19	24.90	24.82	25.64	25.81	26.01	26.60	27.17	27.29	27.17	26.00	25.37
2002	28.55	28.25	28.43	27.77	26.57	25.95	26.53	26.98	26.89	27.24	27.31	27.46	27.33	27.30
2003	26.55	25.76	25.63	26.83	27.89	27.67	28.99	28.59	30.15	29.21	29.70	29.03	28.00	27.51
2004	29.28	27.96	27.76	27.37	26.58	27.12	27.29	26.86	26.81	26.99	26.98	27.62	27.39	27.91
2005	27.39	27.65	27.29	27.68	28.00	28.15	28.54	28.27	27.82	27.63	28.53	28.61	27.96	27.70
2006	28.64	28.79	29.54	29.15	30.72	32.81	34.30	38.36	43.97	40.52	36.22	35.50	34.04	31.75
2007	34.16	33.29	32.67	32.95	32.69	32.17	32.14	31.28	30.13	29.49	27.68	27.87	31.38	33.64
2008	26.91	26.39	26.15	28.04	27.49	27.04	28.65	29.52	28.28	23.77	22.85	22.65	26.48	27.79
2009	21.53	20.45	21.00	24.20										

Source: Servicio Nacional de Informacion de Mercados SNIIM-ECONOMICA.

1/ D.F.- Central de Abasto de Iztapolapa, D.F.

Table 23. Mexico: Sugar production and supply, and sugar and HFCS utilization.

Fiscal Year (Oct/Sept)	1996	1997	1998	1999	2000	2001	2002	2003	2004	2005	2006	2007	2008	2009 1/	2010 1/
							1,000 Metric Tons								
Beginning stocks	1,587	1,403	1,055	991	941	1,063	1,548	1,172	1,194	1,237	1,965	1,294	1,718	1,615	920
Production	4,642	4,818	5,486	4,982	4,979	5,220	5,169	5,229	5,330	6,149	5,604	5,633	5,852	5,550	5,500
Imports	234	191	31	41	37	43	52	63	327	268	240	474	226	400	535
Supply	6,463	6,412	6,572	6,014	5,957	6,326	6,769	6,464	6,851	7,654	7,809	7,401	7,796	7,565	6,955
Disappearance															
Human consumption	4,343	4,301	4,391	4,422	4,445	4,481	5,004	5,097	5,380	5,279	5,326	5,133	5,090	5,340	5,250
Other consumption	71	90	114	127	131	142	180	135	220	282	323	390	414	440	400
Total	4,414	4,391	4,505	4,549	4,576	4,623	5,184	5,232	5,600	5,561	5,649	5,523	5,504	5,780	5,650
Exports	646	966	1,076	524	318	155	413	38	14	128	866	160	677	865	150
Total use	5,060	5,357	5,581	5,073	4,894	4,778	5,597	5,270	5,614	5,689	6,515	5,683	6,181	6,645	5,800
Ending stocks	1,403	1,055	991	941	1,063	1,548	1,172	1,194	1,237	1,965	1,294	1,718	1,615	920	1,155
Stocks-to-human cons.	32.3	24.5	22.6	21.3	23.9	34.5	23.4	23.4	23.0	37.2	24.3	33.5	31.7	17.2	22.0
Stocks-to-use	27.7	19.7	17.8	18.5	21.7	32.4	20.9	22.7	22.0	34.6	19.9	30.2	26.1	13.8	19.9
HFCS cons. (dry weight)	116	402	409	480	580	600	263	130	135	355	667	698	782	610	800

Source: USDA, FAS, PSD database (historical data); USDA, WASDE (forecast data).

1/ Forecast.

E-mail Notification

- Readers of ERS outlook reports have two ways they can receive an e-mail notice about release of reports and associated data.
- Receive timely notification (soon after the report is posted on the web) via USDA's Economics, Statistics and Market Information System (which is housed at Cornell University's Mann Library). Go to
- http://usda.mannlib.cornell.edu/MannUsda/aboutEmailService.do and follow the instructions to receive e-mail notices about ERS, Agricultural Marketing Service, National Agricultural Statistics Service, and World Agricultural Outlook Board products.
- Receive weekly notification (on Friday afternoon) via the ERS website. Go to http://www.ers.usda.gov/Updates/ and follow the instructions to receive notices about ERS outlook reports, Amber Waves magazine, and other reports and data products on specific topics. ERS also offers RSS (really simple syndication) feeds for all ERS products. Go to http://www.ers.usda.gov/rss/ to get started.

End Notes

[1] Beet processors in Michigan source a small portion of their sugarbeets from Canada. Beet sugar from imported sugarbeets is usually less than 1 percent of total production.

INDEX

D

E

F

G

H

I

J

K

L

M

S

T

U

V

W

Y

Z